IRAQ'S NUCLEAR MIRAGE

A map of Iraq with locations mentioned in the book

IRAQ'S NUCLEAR MIRAGE

MEMOIRS AND DELUSIONS

IMAD KHADDURI

SPRINGHEAD PUBLISHERS
Toronto

Copyright © 2003 by Imad Khadduri

First published in 2003 by
Springhead Publishers
22 Springhead Gardens
Richmond Hill, Ontario, L4C 5B9, Canada
Springhead @rogers.com
Telephone 1-905-770-0071
Fax 1-309-419-9281

To order books please contact
Hushion House Publishing Limited
36 Northline Road
Toronto, Ontario M4B 3E2, Canada
Telephone 1-416-285-6100
Fax 1-416-285-1777

National Library of Canada Cataloguing in Publication Data
Khadduri, Imad Y., 1944-
Iraq's nuclear mirage : memoirs and delusions / Imad Y. Khadduri.

ISBN 0-9733790-0-6
1. Khadduri, Imad Y., 1944- 2. Nuclear energy—Iraq—History.
3. Nuclear weapons—Iraq—History. 4. Iraq—History—1958-
5. Scientists—Iraq—Biography. I. Title.

TK9113.I7K49 2003   956.704'3'092   C2003-904980-9

Printed and bound in Canada

TABLE OF CONTENTS

Dedication

Gratitude

My sincere thanks to

Jafar Dhia Jafar,

Dhafir Selbi,
Hussain al-Shahrastani and
Sabah Abdul Noor
(prominent figures in the Iraqi nuclear program)
for their feedback and comments

and

to Peter Kouwenburg for reviewing the manuscript.

INTRODUCTION

This book constitutes the memoirs of an Iraqi who was deeply involved with the Iraqi atomic energy program for thirty years, from its peaceful beginnings to its overtly military applications and its final disintegration.

Through recounting the various scientific roles that I undertook while working for the Iraqi Atomic Energy Commission, the book attempts to reflect the history of Iraq's nuclear program and trace the course that led to the rise of its short-lived potential and ultimately its final demise.

The narrative about the development of the program is combined with the memoirs of one individual who gave the major part of his life to the Iraqi nuclear program. The *"memoirs"* will attempt to give the background to this endeavor, reveal how my commitment to the project was formed, show how tenacious we scientists were in the pursuit of our objectives, and, ultimately, describe my disengagement from the program and departure from my country.

The *"delusions"* of the Anglo-US claims that after the 1991 war Iraqi nuclear weapons existed, or there was still an active program to produce them, have been proven to be part of a massive deception aimed at creating a *casus belli* justifying an invasion of Iraq.

Upon "coming out" in August 2002 when I first realized that President George Bush fully intended to wage a pre-emptive war on Iraq on the pretext that Iraq possessed banned weapons of mass destruction and his emphasis on the nuclear weapons threat, I direly predicted in my first published article, "Iraq's nuclear non-capability", that Bush and British Premier Tony Blair "are pulling their public by the nose, covering their hollow patriotic egging on with once again shoddy Intelligence. But the two parading emperors have no clothes", as is by now quite evident.

The first part of the book exposes the aggressive schemes and uncovers the roots of the American neoconservatives who launched and drove forward the campaign to wage war on Iraq and occupy the country, precipitating its free fall into an abyss.

Furthermore, I contend that the neoconservatives—acting in the service of Israel's interests—adroitly exploited the inherent cruelty and insecurity in the "American Way of Life" to attain their ends. I recognized and experienced the cruel and insecure aspects in the "American Way of Life" during my university years (1961-68) in the US.

Lies and distortions not only led to the occupation of Iraq, but also, between 1990 and 2003, enabled the US and Britain to maintain the punitive sanctions regime which caused the deaths of hundreds of thousands of Iraqi civilians. Since there was no evidence that Iraq retained stockpiles of weapons of mass destruction after the early nineties, the strangulating sanctions regime should have been lifted long ago. The fact that the sanctions were stubbornly maintained by these two powers is a monstrous crime which makes the thousands of deaths caused by the war of occupation pale in comparison.

By the time they waged this war, an incapacitated and heavily damaged Iraq was like a collection of beads strung precariously together by a notoriously repressive dictator. The US, in its reckless bravado cut the string.

Today, the US is, all at once, attempting to implement multiple objectives in Iraq: eliminate weapons of mass destruction it cannot find, combat terrorism, remove all traces of the Saddam regime, establish a free economy, build a democracy and normalize relations between Iraq and Israel. In reality, the US cannot fill the vacuum it created by abolishing the 80-year old Iraqi state. In fact, the US appears to be stuck in its tracks like a wet hand on a frozen metal. Its ruthless historical rise to political preeminence does not mean it has the capability to comprehend or repair the damage it has wrought unto Iraq.

However, the Iraqi people will rebuild their country.

August 2003

CHAPTER ONE:

The Rape of Strangled Iraq

March 20, 2003 will be as infamous to the people of Iraq as September 11, 2001 is for the American people. On these days, both societies felt the shock of mass terror. While US moral credibility, in my opinion, had reached its zenith on September 10, Iraq hurtled toward a nadir, yet to be reached, but will, nonetheless, rebound.

However, Iraq's society itself is being ruthlessly raped. A deliberate strangulation had preceded the pillage. The United States of America had vehemently imposed, through the United Nations, an economic embargo as barbaric as a medieval siege. For thirteen years, the sanctions wantonly destroyed Iraq's civilian infrastructure, such as electric power plants, water and sewage plants, education, the health sector and economy. This produced a degree of suffering on the Iraqi people that was never fully comprehended in the West, the news of which adroitly filtered by their media. Documented evidence was available, volumes of it. Yet back in 1996, when the number of Iraqi children killed off by sanctions stood at around half a million, Secretary of State Madeleine Albright made her infamous declaration to Lesley Stahl on CBS that "we think the price is worth it".

Two senior United Nations officials responsible for the humanitarian relief in Iraq, Denis Halliday and Hans von Sponeck, who had resigned in the late nineties, protested the embargo's hidden agenda. Halliday expressed a widely held belief when he said in 1998, "We are in the process of destroying an entire society. It is as simple and terrifying as that." He called it "genocide".

Compounded by a ruthless, self-aggrandizing despot who ruled with an ever increasing repression that was perpetrated by about eighteen Security and Intelligence organizations, the Iraqi people suffered immensely. It would be rare to find an Iraqi extended family who have not seen at least a loved one executed, die in a war or imprisoned during Saddam's thirty years of ruthless reign of terror, including his own.

These two repressive factors, in addition to two wars, in the eighties and in 1991, had destroyed Iraq's infrastructure, squandered the accumulated experience of its nation building that was laboriously catered for by its professional class since Iraq's statehood in 1921 and shredded its social, educational, legal and economic fabric.

Coincidently, a new ruthless force was emerging in the United States of America that has spearheaded the bloody rape of Iraq.

The neoconservatives: AEI and PNAC

As a result of several bizarre and unforeseeable contingencies, such as the selection rather than the election of Bush and the events of September 11, the foreign policy of the world's only global power has come to be dictated by a small clique that is unrepresentative of either the US population or the mainstream American foreign policy establishment but had managed to move to the centre of President Bush's foreign policy agenda.

The core of the group that is now in charge of these policies consists of neoconservative defense intellectuals. They are called "neoconservatives", or neocons, because many of them started as anti-Stalinist leftists or liberals before moving to the far right. Inside the US government, the chief defense intellectuals include Paul Wolfowitz, the deputy secretary of defense and Donald Rumsfeld who holds the position of Defense Secretary only because Wolfowitz himself is too controversial. Others include Douglas Feith, the Under Secretary of Defense and Policy Advisor at the Pentagon; Lewis Libby, a Wolfowitz protégé who is Vice President Cheney's chief of staff; John R. Bolton, a right-winger assigned to the State Department to keep Colin Powell in check; and Elliott Abrams, recently appointed to head Middle East policy at the National Security Council. Outside the government are James Woolsey, the former CIA director, who has tried repeatedly to link both the September 11 and the anthrax letters in the US to Iraq, and Richard Perle, who has recently resigned his unpaid chairmanship of a defense department advisory body after a lobbying scandal. Most of these "experts" have never served in the military. Their headquarters is now the civilian Defense Secretary's office, where these Republican political appointees are despised and distrusted by the largely Republican career military generals.

In his elucidating article[1], Michael Lind demarks the neocon's territory of influence. The neocon defense intellectuals, as well as being in or around the actual Pentagon, are also at the center of a metaphorical "pentagon" of conservative think tanks, the Israeli lobby, religious right groups and media empires.

Think tanks such as the American Enterprise Institute (AEI) provide resting homes for the neocons when they are not currently serving in the government. The funding comes not so much from industrial corporations as from decades-old conservative foundations, such as the Bradley and Olin foundations, which spend from the estates of long-dead tycoons. Neoconservative foreign policy does not reflect business interests overtly, though they are directly nourished

---

[1] *"How neo-conservatives conquered Washington—and launched a war"*, by Michael Lind, Whitehead Fellow at the New America Foundation in Washington. April 9, 2003.
http://www1.iraqwar.ru/iraq-read_article.php?articleId=2334&lang=en

by it. "The neocons are ideologues, not opportunists".

The major link between the conservative think tanks and the Israeli lobby in the US is the Washington-based and Israeli Likud-supporting Jewish Institute for National Security Affairs (JINSA), which has already co-opted many non-Jewish defense experts by sending them on trips to Israel. It included, for example, the retired general Jay Garner, the failed first proconsul of occupied Iraq. In October 2000, he had cosigned a JINSA letter that began: "We... believe that during the current upheavals in Israel, the Israel Defense Forces have exercised remarkable restraint in the face of lethal violence orchestrated by the leadership of [the] Palestinian Authority[1]."

The Israel lobby in the US itself is divided into Jewish and Christian wings. Wolfowitz and Feith have close ties to the Jewish-American Israel lobby. Wolfowitz, who has relatives in Israel, has served as the Bush administration's liaison to the American Israel Public Affairs Committee. Feith was given an award by the Zionist Organization of America, citing him as a "pro-Israel activist". While out of power in the Clinton years, Feith collaborated with Perle in 1996 to coauthor a policy paper entitled "A Clean Break: a New Strategy for Securing the Realm"[2]. The realm is the Israeli one in the Middle East and it was prepared for the incoming Israeli government of Binyamin Netanyahu. It examined various ways in which Israel could "shape its strategic environment", beginning with the removal of Saddam Hussein and the installation of a monarchy in Baghdad. It also advised the Israeli government

---

[2] *"Right takes centre stage"*, The Guardian, by Brian Whitaker, March 4, 2003. http://www.guardian.co.uk/elsewhere/journalist/story/0,7792,907312,00.html Referencing *"A Clean Break: A New Strategy for securing the Realm"*, a report prepared by The Institute for Advanced Strategic and Political Studies' "Study Group on a New Israeli Strategy Toward 2000." The main substantive ideas in this paper emerge from a discussion in which prominent opinion makers, including Richard Perle, James Colbert, Charles Fairbanks, Jr., Douglas Feith, Robert Loewenberg, David Wurmser, and Meyrav Wurmser participated. http://www.israeleconomy.org/strat1.htm

to end the Oslo peace process, reoccupy the territories, and crush Yasser Arafat's government. These goals are clearly discerned in the present overtly aggressive reign of Ariel Sharon.

Such experts are not typical of Jewish-Americans, who had mostly voted for Al Gore in 2000. The most fervent supporters of Likud in the Republican electorate are Southern Protestant fundamentalists. These religious right groups believe that God gave all of Palestine to the Jews, and fundamentalist congregations spend millions of dollars to subsidize Jewish settlements in the occupied territories.

The final corner of the neoconservative "pentagon" is occupied by several right-wing media empires, with roots in the British Commonwealth and South Korea. Rupert Murdoch disseminates propaganda through his Fox television network. His magazine, the Weekly Standard, edited by William Kristol (Chairman of the Project for the New American Century), acts as a mouthpiece for defense intellectuals such as Perle, Wolfowitz, Feith and Woolsey as well as for Sharon's government. The National Interest is now funded by Conrad Black, who owns the Jerusalem Post and the Hollinger Empire in Britain and Canada.

Strangest of all is the media network centered on the Washington Times, owned by the South Korean messiah (and ex-convict) the Rev. Sun Myung Moon, which owns the newswire UPI. UPI is now run by John O'Sullivan, the ghostwriter for Margaret Thatcher who once worked as an editor for Conrad Black, the media czar in Canada. Through such media channels, the shallow and sensational style of right-wing British tabloid journalism has contaminated most of the US news media.

How did the neocon defense intellectuals, a small group at odds with most of the US foreign policy elite, Republican as well as Democratic, manage to capture the Bush administration? Few of them had supported Bush during the presidential primaries. They feared that the second Bush would be like the first, a wimp who had failed to occupy Baghdad in the first Gulf War and who had pressured Israel into the Oslo peace process, and that his administration, again like his father's, would be dominated by moderate Republican realists such as Powell, James Baker and Brent Scowcroft. They did support the maverick senator John McCain until it had become clear that Bush would be nominated.

Then they had a stroke of luck. Cheney was put in charge of the presidential transitional period (the period between the election in November and the accession to office in January). Cheney exploited this opportunity to stack the administration with his hard-line allies. Instead of becoming the de facto president in foreign policy, as many had expected, Secretary of State Powell found himself boxed in by Cheney's right-wing network, including Wolfowitz, Perle, Feith, Bolton (Undersecretary of State) and Libby (Vice President Dick Cheney's Chief of Staff).

The neocons took advantage of Bush's ignorance and inexperience. Unlike his father, a Second World War veteran who had been ambassador to China, director of the CIA, and vice president, Bush was a thinly educated playboy who had failed repeatedly in business before becoming the governor of Texas, a largely ceremonial position, with questionable military service record. His father is essentially a northeastern moderate Republican. Bush, who was raised in west Texas, had absorbed the Texan cultural combination of machismo, anti-intellectualism and overt religiosity. The son of upper-class Episcopalian parents, he converted to Southern fundamentalism in a midlife crisis. Fervent religious connotations are reflected in many public utterances of Bush, the most glaring of which was reported in Ha'aretz[3], Israel's most respect-

---

[3] *"Road map is a life saver for us," PM Abbas tells Hamas"*, by Arnon Regular, Ha'aretz, July 26, 2003.
http://www.haaretzdaily.com/hasen/pages/ShArt.jhtml?itemNo=310788&contrassID=2&subContrassID=1&sbSubContrassID=0&listSrc=Y

ed daily. In it, the Palestinian prime minister, Mahmoud Abbas, told the Israeli newspaper that Bush made the following pronouncement during a meeting between the two: "God told me to strike at al-Qaeda and I struck them, and then He instructed me to strike at Saddam, which I did, and now I am determined to solve the problem in the Middle East."

The osmosis of neocon thought had already infected the newly ordained Bush. A blueprint for US global domination reveals that Bush's future cabinet was planning a premeditated attack on Iraq to secure a 'regime change' even before he took power in January 2001[4].

The blueprint for the creation of a 'global Pax Americana' had been drawn up in September 2000 for Dick Cheney, Donald Rumsfeld, Paul Wolfowitz, Bush's younger brother Jeb and Lewis Libby. The document, entitled "Rebuilding America's Defenses: Strategies, Forces and Resources for a New Century"[5], was written by the neoconservative think-tank Project for the New American Century (PNAC) that was founded in 1997.

The plan shows Bush's future cabinet had intended to take military control of the Gulf region whether or not Saddam Hussein was in power. It claims that, "The United States has for decades sought to play a more permanent role in Gulf regional security. While the unresolved conflict with Iraq provides the immediate justification, the need for a substantial American force presence in the Gulf transcends the issue of the regime of Saddam Hussein".

The PNAC document supports a "blueprint for maintaining global US pre-eminence, precluding the rise of a

---

[4] *"Bush Planned Iraq 'Regime Change' Before Becoming President"*, By Neil Mackay
http://www.informationclearinghouse.info/article1221.htm

[5] *Rebuilding America's Defenses, "Strategy, Forces and Resources For a New Century"*, A Report of The Project for the New American Century, September 2000.
http://www.informationclearinghouse.info/pdf/RebuildingAmericasDefenses.pdf

great power rival, and shaping the international security order in line with American principles and interests".

This "American grand strategy" must be advanced for "as far into the future as possible", the report says. It also calls for the US to "fight and decisively win multiple, simultaneous major theatre wars" as a "core mission".

Bush was tilting away from Powell and toward Wolfowitz even before September 11 had given him something he had lacked: a mission in life other than following in his dad's footsteps. There were signs of estrangement between the cautious father and the crusading son. In 1992, veterans of the first Bush administration, including Baker, Scowcroft and Lawrence Eagleburger, had warned publicly against an invasion of Iraq without authorization from the American Congress and the United Nations.

It is not clear if Bush fully understands the grand strategy that Wolfowitz and other aides are unfolding. Aside from his newly embraced doctrine of pre-emptive wars, he seems to genuinely believe that there was an imminent threat to the US from Saddam Hussein's "weapons of mass destruction", something the leading neocons say in public but, one would assume, should be far too intelligent to believe themselves. The Project for the New American Century had urged an invasion of Iraq throughout the Clinton years, for reasons that had nothing to do with possible links between Saddam and Osama bin Laden. Public letters signed by Wolfowitz and others called on the US to invade and occupy Iraq, to bomb Hezbollah bases in Lebanon, and to threaten states such as Syria and Iran with US attacks if they continued to sponsor terrorism. Claims that the purpose is not to protect the American people but to make the Middle East safe for Israel are glibly dismissed by the neocons as vicious anti-Semitism.

Thus have neoconservative dogmas taken over Washington and steered the US into a burgeoning Middle Eastern war starting with the occupation of Iraq. This was unrelated to any plausible threat to the US and opposed by the public of every country in the world except Israel. The frightening thing is the role of happenstance and personality. After the al-Qaida attacks, any US president would likely have gone to war to topple bin Laden's Taliban protectors in Afghanistan. However, everything that the US has done since then would have been different had America's 18th century electoral rules not given Bush the presidency and had Cheney not used the transition period to turn the foreign policy executive into a PNAC reunion.

A more distant perspective of the neoconservatives' thinking is noteworthy.

Neoconservative roots

Albert Wohlstetter

In her detailed exposition "The Neocons in Power"[6], Elizabeth Drew traces the intellectual roots of Perle, Woolsey, and Wolfowitz to the late Albert Wohlstetter, a University of Chicago professor who had worked for the RAND Corporation and later taught at the University of California. Throughout the cold war, he had argued that nuclear deterrence was not sufficient and that the US had to actually plan to fight a nuclear war in order to deter it. He strongly advocated the view that the military power of the USSR was underrated. Wolfowitz earned his Ph.D. under Wohlstetter; Perle met Wohlstetter when he was a high school student in Los Angeles and was invited by Wohlstetter's daughter to swim in their pool. Later, Wohlstetter invited Perle, then a graduate student at Princeton, to Washington to work with Wolfowitz on a paper about the proposed Anti-Ballistic Missile Treaty, which

---

[6] *"The Neocons in Power"*, by Elizabeth Drew. The New York Review of Books, VOLUME 50, NUMBER 10, JUNE 12, 2003
http://www.nybooks.com/articles/16378

Wohlstetter opposed and which has been abandoned by the senior Bush administration. Ahmed al-Chalabi met Wohlstetter while he studied for a doctorate in mathematics, at the University of Chicago.

Through Wohlstetter, Perle had also met Ahmed al-Chalabi. Ahmed had by then founded the Iraqi National Congress in 1992, an umbrella organization of Iraqi opposition groups, many of its members in exile and overtly funded by the CIA. These exiles had proved to be the conduit for an abundance of false Intelligence information. This dubious information was the fodder for the neoconservatives' misinformation campaign that sustained the march to the occupation of Iraq and that will yet prove to be the neocons' undoing. Perle had invited Ahmed al-Chalabi to an annual conference in Beaver Creek, Colorado, which was cosponsored by the American Enterprise Institute, and former president Gerald Ford, as his guests. At such conferences, al-Chalabi was able to meet Cheney, Rumsfeld, and Wolfowitz and secure their backing.

Michael Ledeen
A fellow at the conservative American Enterprise Institute, Michael Ledeen has become the driving philosophical force behind the neoconservative movement and the military actions it has spawned[7]. His 1996 book, "Freedom Betrayed; How the United States Led a Global Democratic Revolution, Won the Cold War and Walked Away," reveals the basic neoconservative obsession: the United States never "won" the Cold War; the Soviet Union collapsed of its own weight without a shot being fired. Had the United States truly won, democratic institutions would have been sprouting wherever the threat of communism had been rife.

---

[7] *"Military Might, The man behind 'total-war' in the Middle East"*, sfgate.com, William O. Beeman, May 14, 2003.

Iraq, Iran and Syria are the first and foremost nations where this should be happening, according to Ledeen. The process by which this should be achieved is a violent one, termed "total-war".

"Total-war not only destroys the enemy's military forces, but also brings the enemy society to an extremely personal point of decision, so that they are willing to accept a reversal of the cultural trends," Ledeen writes.

"The sparing of civilian lives cannot be the total war's first priority.... The purpose of total-war is to permanently force your will onto another people."

The journalist Thomas Friedman, on a tour of Iraq wrote in the New York Times in May 25, 2003, "The best thing about this poverty: Iraqis are so beaten down that a vast majority clearly seem ready to give the Americans a chance to make this a better place." Ledeen depravedly expects to be vindicated. The Iraqi people will prove him otherwise.

Ledeen's ideas are repeated daily by such figures as Cheney, Rumsfeld and Wolfowitz. His views virtually define the stark departure from American foreign policy philosophy that existed before the tragedy of September 11, 2001. He basically believes that violence in the service of the spread of democracy is America's manifest destiny. Ledeen has become the philosophical legitimizer of the American occupation of Iraq and the wanton destruction of its society.

After using the "good offices" of UN diplomacy (economic sanctions and weapons inspections) to ensure that Iraq was brought to its knees, its people starved, half a million children unnecessarily dead, its infrastructure severely damaged and after making sure that most of its weapons had been destroyed, the "Coalition of the Willing", more apt to be called as the Coalition of the Bullied and Bought, sent in a technologically superior army. The gallantry of its soldiers, not bravely claimed from thousands of feet in the air, is thoroughly being exposed to what it really is on the ground.

Vandals plundered museums, ministries, offices, hotels and hospitals. American and British soldiers stood by and watched. They said they had no orders to act. In effect, they had orders to kill people, but not to protect them. Their priorities were clear. The safety and security of Iraqi people was not their business. The security of whatever little remained of Iraq's infrastructure was not their business. However, the security and safety of Iraq's oil fields were. The oil fields were "secured" even before the invasion began. The Oil Ministry was heavily protected immediately afterwards.

This was followed by what analysts call a "power vacuum". Cities that had been under siege, without food, water, and electricity for days and relentlessly bombed, with millions of people who had been starved and systematically impoverished by the UN sanctions for more than a decade, were suddenly left with no semblance of security, control or urban administration. A seven-thousand-year-old civilization slid into chaos. Rumsfeld's anarchistic comments on this were astounding, "It's untidy. Freedom's untidy and free people are free to commit crimes and make mistakes and do bad things." American TV channels lingered on the few jubilant Iraqis whose hate for Saddam Hussein blinded them from seeing the fact that they had just fallen into the net of another headhunter, who cared little while dropping bombs on civilian areas so that it might confidently declare, "we have liberated 600 oil fields"[8].

The neoconservatives have come to believe that their intellectual superiority entitles them to rule over the bulk of humanity by means of duplicity. Public support in the US for the war against Iraq was founded on a multi-tiered edifice of falsehood and deceit, coordinated by the US govern-

---

[8]   *"What Else Hasn't Israel Told America?"*, by Ramzi Baroud, The Palestine Chronicle, April 21, 2003.
http://www.palestinechronicle.com/article.php?story=20030421081806395

ment and faithfully amplified by the corporate media[9].

Horrendously, an ancient civilization is casually being decimated by a very recent, casually brutal nation.

Just as I have faith that the Iraqi people will resurrect from their calamity, I also believe that the American people will come to realize the folly the neoconservatives have herded them in and seek retribution.

Herein begins my memoirs. It attempts to expound the background of my preparedness for, the commitment to and the involvement in the Iraqi nuclear program, from its beginnings to its demise. It concludes with my humble efforts in attempting to expose the misinformation campaign that was wielded in support of the occupation of Iraq.

The path is still not yet fully traversed.

---

[9]   "*The Day of the Jackals*", a speech at a national-antiwar teach-in by Arundhati Roy, June 2, 2003.

CHAPTER TWO:
Growing Arabic

A solid childhood

At the end of 1944, Baghdad was a kaleidoscope of a city to be born in. Bedouin tribal affiliations collided with nouveau riche professional middle class values yet enshrouded with a spectrum of religions that reflected Iraq's rich history and colorful traditions.

My father, Yousif Yaqoub Khadduri, was a medical doctor who had just finished his military service. It had been only 25 years since the end of World War I, with a practically non-existent professional class in Iraq, when it gained its so-called "independence" after three centuries of Ottoman rule. The British had also come as "Liberators" only to later become liberated from Iraq.

My grandfather was a wealthy merchant in the southern port city of Basra. In 1929, he partnered with a fellow Christian merchant from the northern city of Mosul and sent to the US a ship laden with Iraqi leather collected from most of the northern and southern Arab tribes for that year, to be paid for when the goods were sold. The ship arrived at New Orleans during the Stock Market crash in 1929 and sold for a pittance. Honoring their words, the two merchants had to compensate the tribesmen for their goods. Declaring bankruptcy was not a morally accepted escape route. My grandfather sold all his property and possessions. A year later, on his deathbed, he had asked his brother, Toma, for a favor. My grandfather had always intended that Yousif and Khadduri, his two sons, should seek higher education. He obliged Toma to see to that. My father and uncle were sent, at Toma's expense, to the American University of Beirut in the following year. In 1934, they returned to Baghdad, my father a medical doctor and my uncle a lawyer.

Despite her easily discerned intelligence, my mother, Mari Abaji, remained a housewife all her life. Throughout her life, she never forgot her perfectly fluent French that she had acquired at the famed Rahibat Christian nuns' primary and secondary school in Baghdad. Her father's family was from Aleppo, Syria. He was also a merchant, with a penchant for beautiful women. He married into the Bazzoui, one of the firmly established Christian families of Iraq.

Both of my parents had lived their youth in Aqid al-Nassara, the Street of the Christians, in the middle of Baghdad, where the alleys would just allow for two laden donkeys to go through side by side. The balcony of the houses nearly touched at the upper levels and the sun rarely made it through the cracks of architecture to reach the alley. All the houses had cellars where everybody retired during the hot noon hours of the summer, and where scorpions also sought some cool solace. The roofs were wetted with water in the evening to dampen the atmosphere at night before everybody slept on the roof with their water clay jugs, barely sheltered from the neighbors. It still allowed pigeon enthusiasts to peek through at sleeping lasses, as they pretended to fly their home raised flock of pigeons. That was considered so immoral that these pigeon caretakers were shunned from having the privilege of giving testimony in a court of law.

My family, along with Walid, my elder brother, had moved to Bistan al-Khas, the Lettuce Garden, which was on the outskirts of the tightly bound Baghdad, ten kilometers by ten kilometers at the time, and divided by the majestic, often deadly flooding Tigris River. Gozi al-Qas, a beautiful and benevolent Assyrian lady from the ancient city of Tilkaif in the north, was our nanny.

I was, I am told, a hot pepper of a child. Gozi recounted many of my escapades when I often visited her forty years later.

My mother could not keep up with my enthusiasm and pranks. She begged Madam Adil, the headmistress of Adil primary school, where my father was the medical doctor for the school, to accept me into kindergarten even though I was still only three years old, below the legal admittance age of four years. Madam Adil had me repeat the second kindergarten level so that I would be of legal age for entrance to primary school. I would still live in boarding at Madam Adil for several years due to the hyperactive level of my childish exuberance.

Madam Adil, as the Lebanese Zihoor Adil was lovingly called, was a no nonsense head mistress. She and her husband, Dr. Anis Adil, had come to Baghdad from Lebanon and established the Adil Private School in 1933. She was such a wonderfully dominating head mistress that the school itself was affectionately known simply as Madam Adil. The school had a galaxy of excellent teachers. It was one of the few best primary schools in Baghdad. My classmates came from the top professional and merchant Moslem, Christian and Jewish families of Baghdad. Its discipline and quality of education ensured that its graduating sixth grade class collected the top marks in the yearly government exam that was given to all primary schools in Iraq. Ahmad al-Chalabi, who had become known in 2003 for his dubious role in the occupation of Iraq, was in my graduating class in 1956. Madam Adil sadly closed after the 1958 revolution in Iraq. Most of the boys from Madam Adil went straight to Baghdad College, the most elite high school in Baghdad.

The American Jesuit Fathers had established Baghdad College in 1932. Run by a highly motivated and professional staff of Jesuit Fathers, it soon became the standard for quality education throughout the Middle East. In order to prepare its students for the two government examinations that were administered throughout Iraq at the third and fifth year of secondary education, the students had to take each course, except for Arabic language, both in Arabic and in

English. The textbooks and laboratories were consistently up to date. Extracurricular activities included debating, elocution, drama, and athletics. Being a Catholic, I had to be immersed in a heavy dose of Christian faith indoctrination, only to rebel afterwards during my university education. The educational standard was so high that only about 100 students out of the initial batch of 250 students that entered Baghdad College every year would manage to graduate after the five grueling years of its secondary education. The Jesuit Fathers also opened and ran al-Hikma University from 1956 until their ouster by the Ba'athists from Iraq in 1969.

Political consciousness was rife, to the chagrin of the Jesuit Fathers. I was inducted as a supporter to the Ba'ath party at the ripe age of 14, while still at second grade in high school. My mentor was Adil Abid al-Mahdi, of emerging political fame as a leading official of the largest Shiite group, the Supreme Council of the Islamic Revolution in Iraq, which is headed by Ayatollah Muhammad Bakir al-Hakim, now exercising its power openly in Iraq after the 2003 occupation.

We participated in many demonstrations, some violent, against the police of the repressive Nuri al-Said, the perennial prime minister in the Kingdom of Iraq and later Abdul Karim Qassim, the President of the later Republic. One of the most serious confrontations against the Jesuit Fathers occurred in 1961, the year of my graduation and after we had taken the final exams. I had already taken the Test of English as a Foreign Language (TOEFL) exam and applied to several distinguished American universities. The police had just picked up my best friend, the late Hashim Abid al-Mahdi, the brother of Adil Abid al-Mahdi, my Ba'athist mentor. Hashim's sister phoned, warning me that she had heard the police mention that they were coming to pick me up next, just down the road from Hashim's home. My father drove me to the railroad station, and gave me enough money to flee to the resort of Salah al-Din in the North. There I met

Imad Khadduri, Basil al-Qaisi and Dhafir Selbi in Salah al-Din, 1961.

my erstwhile friends, Basil al-Qaisi, who later inducted me in the Iraqi nuclear program, and Dhafir Selbi, a prominent key figure in Iraq's nuclear weapons program and my boss therein.

At Salah al-Din, I was distressed by receiving the mail sent by my father informing me of the flat rejection of my university applications by most of the American universities that I had applied to, without mentioning the reason why. My last hope was Michigan State University (MSU) where my brother Walid had already been attending for two years. They finally also sent me their rejection, but this time explaining the reason. I had done poorly in the English section of TOEFL while excelling in all other sections of that test. The fact descended upon me as to why I had been rejected by all universities. The TOEFL that we had taken in Iraq was administered wrongly to us. It was intended for American students, whose mother tongue was English, and not for foreign students; hence my poor showing in that particular section despite our solid English schooling. I wrote a letter to the Registrar at MSU explaining this dichotomy and pointing to the success of my brother at MSU, while brandishing the fact that my high school grades were substantially higher than his. Within three weeks, I received the acceptance from Michigan State University. Later that Christmas, I was working as a file clerk for the administration in MSU, and I stumbled upon the letter that I had sent them from my mountain hideout. The Registrar had scribbled on it: "If he can write such good English, then accept him".

While still hiding at Salah al-Din in the summer of 1961, I was ordered by Adil to negotiate with the Jesuit Fathers. I returned to Baghdad and visited them. I had to smuggle myself in the trunk of a taxi since the police had surrounded the grounds of Baghdad College, and did manage to arrive at a mutually agreeable solution to the outstanding conflict with the Jesuit Fathers.

Due to my high-grade average in the government examination at the end of secondary school, I was offered a nine-year government paid scholarship to pursue a PhD program in nuclear physics in the United States of America. Such was the commendable state of support, since the thirties, by the Iraqi government to higher education. However, my father turned down the offer. He wanted us to be free from the bondage of the scholarship, which stipulated that, upon graduating, we should serve double the expended time working at an Iraqi government Ministry. He chose to support the higher education of both his sons, regardless of cost. It was a heavy burden on him.

A few weeks later, at the age of seventeen, I left Iraq to East Lansing, Michigan to attend Michigan State University.

CHAPTER THREE:

Alienation and Fear

An open mind in the belly of the beast

The wide highways and speeding cars most impressed me when I first landed at New York in late August 1961. My older brother, Walid, quickly hauled me to East Shaw dormitory at Michigan State University in East Lansing, Michigan, since the Fall Term was about to begin. I woke up in the middle of my first night there with an acute case of appendicitis, and was rushed to the hospital. The surgeon confided to Walid of his surprise at the strength of my abdominal muscles.

As a testimony to the solid educational quality imparted to us at Baghdad College, my first term grades at MSU amounted to a 4.0 grade point average, or an A+ for a full load of courses. The President of MSU held a dinner invitation for all of the 4.0 grade point average students at the Student Union. We were arranged along long rows in a huge hall. I was at the last row from the President. He gave an excellent speech, closing it with his trademark slowly winding crescendo, in reverse. He then asked us a favor, meaningless and mildly perturbing to my still unpolished grasp of stage show etiquette. Would each of us stand up and recite our name and where we came from? There were about 250 of us. As the name annunciation slowly snaked its way to my last row, I had had close to enough of it. I stood up reciting my name and my origin. Sitting down, the pangs of subdued displeasure busted. I jumped up, waving the fork that was still in my grasp, and announced, "My student number is 336458", and sat down relieved. I was ridiculing the triviality of the ritual. A meaningless, faceless student number would be as good as listening to 250 names and "I come from…" I was surprised to hear some applause, after a second of stunned silence.

Melanie Bach was the first woman with whom I shared an intimate experience. Her open desires were often painfully thwarted by my strict Catholic bondage and fear of religious retribution that held me back. Her frustrated tears reflected her amazement at my cling to this orthodoxy and her disbelief that it was not her lack of attractiveness that distanced me from her body. Our relationship furthered my departure from Catholicism.

During that first winter at Michigan, I joined the Catholic Club to a trip to a ski resort. There were four of us, all males, in an old convertible. The whole party drove to the destination in pairs of cars that tagged each other. A snow blizzard was blowing and the two cars trailed each other carefully. Unfortunately, we had a flat tire and the other car helped us in replacing it. An hour later, and as the other car impatiently pulled ahead despairing of our slow progress, we had a second flat tire and no spare one. It was about eleven at night and the highway was deserted with hard blowing snow and wind that streaked through the cracks and the tears of our car's convertible top. The car's heater was not functioning. The temperature outside was about minus 25 degrees centigrade. Not a car passed by. Agonizing hours did. It did not take long for our limbs to feel numb. I shared my packets of dates, for their supposedly inherent heat source, with my colleagues. At about two o'clock in the morning, our driver began to doze off into unconsciousness. With the help of the others, we moved him to the side chair and I sat behind the wheel, praying and hoping for a car to pass by. Finally, headlights lit up our rear view mirror. I barely managed to flash our headlights, with my nearly useless numb fingers. It was a truck, and by the time it stopped, it was about fifty meters ahead of us. I could not move to open the door and signal them to come for help. The truck drivers remained in their truck, peering in their mirror expecting a response. At that point, a car, heading down the other direction stopped across from us. Our partner car had arrived at the destination, missed our arrival, and decided to turn back and search for us. They and the truck drivers had to singly carry each one of us in frozen postures from our convertible and into the other warm car. The sharp pain of frostbite reminded me of the pain from a tonsil operation after the morphine faded off.

My father was struggling hard to maintain his promise to pay the university tuition and lodging for both my brother and myself. Many years later, we learned of how difficult it was for him, having even been forced to sell the only piece of prime land that was assigned to him by the government, by virtue of his being a medical doctor, and swallowing his pride and ask for a loan from a relative. My parents did not manage to own a home until after our graduation. He had only 500 Iraqi Dinars ($150) to his name upon his passing away in 1989.

At the end of my first year at East Shaw dormitory, I applied for and got a job as a Resident Assistant (RA). The position called for managing internal dorm discipline, athletic activities and general counseling for about 50 students. There were about 12 other RAs in the dormitory. Our compensation was free room and board for the academic year. The assignment continued for three years until my graduation in 1965.

The first cracks in my pristine image of the "American Way of Life" started to form during my RA experience. Being nurtured upon strong friendship bonds, trusts and loyalties while growing up in Baghdad, it struck me that something was really amiss since after three years' experience as an RA, living with and supervising more than 150 American students, I could not discern even a single true

Resident Assistant at East Shaw Hall, Michigan State University, 1963.

friendship relationship among them; at least, not the kind of friendship that I had valued, yearned for and had previously felt and cherished. The relationships that I witnessed, during these formative and still young age, appeared to me as shallow and easily disposable. They were not, I felt, the kind that you would recall many years later and claim that these two people were friends. The prevalent relationship was akin to acquaintances, easily unstuck and faded. Looking back at that conviction four decades later, I would tend to temper that distinction as not being very accurate.

On top of the study load, I also cast a wide net of reading, such as the books of the individualist Ayn Rand, the revolutionary visionaries Regis Debre and Che Guevara and the true revolutionaries Karl Marx, Lenin and Mao Tse Tung. Then there were some mind shattering books on "Alienation". At the time, such a word was not even translated into Arabic, being such an alien concept to the Arabic psyche. It was, and still is, extremely difficult to explain the meaning of the term to an Arab immersed in his social milieu, to have him feel the transplantation of human relationships and considerations with materialistic and cold calculations. Combined with heightened capitalistic conditions, it seemed that many in the US were truly becoming "hollow" and "empty" in the words of the poet T. S. Eliot.

During the summers of the years 1964, 1965 and 1966, I would purchase an inexpensive round trip airplane ticket to London. From there, I would spend three months hitchhiking throughout Europe with the aid of a pen and large sheets of paper on which I would scribble "Student" and the name of the next city that I wanted to travel to. My travels covered all European countries. Titillating experiences abounded; watching Nureyev and Fontaine dance in Florence accompanied by a Scottish lass who journeyed with me further, skinny-dipping in Danish lakes and staying at posh mansions in Germany with Miele's heiress whom I had met earlier at Michigan State University. These travels would invariably end in Beirut, Lebanon where I was an appreciative guest of Jack Bazzoui, a distant bachelor relative of mine, who had a taste for the good life and its pleasures.

Even though these summer excursions would cost me $200-$300 per summer (since lodging would be free with newly found friends or cheap at student hostels), I did, on one occasion, end up in Bari, in southern Italy, without a fare for a voyage on a ship to Athens, where a modest bank transfer from my father would have been waiting for me. The travel agent would not be appeased by my assurances that he would be repaid once I got to Athens. At loss, I noticed a few cars with American plate numbers. They were from the nearby American military base, explained the travel agent. I whipped my pen and scribbled on an empty paper sheet: "Please stop. I need help not a ride". An American officer screeched his convertible to a stop. He was a fellow University of Michigan alumnus, and by virtue of that, we went back to his apartment for a beer, a steak and a first class boat ticket to Athens. First, he asked me a favor. He had ordered a Tiger's Eye ring from a jeweler in Beirut for his fiancé, which had not arrived even though he had paid for it. Picking up my first class ticket wistfully from the wide-eyed travel agent, I made sure that that ring was dispatched faithfully from Beirut.

The dark clouds of the Vietnam War had by then turned into a storm. A book resonated with my comprehension and feelings on what was befalling the United States of America and pointed my awareness to the understanding that I was groping to formulate. The 1958 novel, "The Ugly American" by William Ledere and Eugene Burdick, helped bring into focus the disparate questions, impressions, puzzles, feelings and antipathy that were swirling in my head. The outlines of the beast, from within his belly, were forming. This can be seen by the following event.

As the crescendo against the war gathered momentum, Dow Chemical Company saw it fit to give a presentation to the University of Michigan students to defend their production of napalm that was being used extensively in Vietnam. It occurred after the Detroit summer riots of 1967 when tanks were used in the streets of Detroit to quell the disturbances. The Dow Chemical representative gave a polished defense of his company to the hundreds of assembled students. I asked two related questions. Does Dow support the use of napalm by the American forces against civilians in Vietnam? A militaristic justification was presented by Dow's spokesman. I then asked the rhetorical question. Suppose that matters did get out of hand in Detroit, and that tanks were not enough to control the rioting, burning and looting; and the army had decided to use napalm to control the situation. Would Dow sanction the use of napalm against the black people in Detroit? No, was the response, after a moment's reflection. Then how does Dow justify the use of this horrific indiscriminate weapon against civilians in Vietnam and not in the US? Is there not a double standard as to the worth of human beings postulated here? There was no answer, but some applause.

In the mid-sixties, the daunting distaste for the "American Way of Life", coupled with my heightened appreciation of the richness and warmth of Arabic culture, led me to take a fateful decision. I would only marry an Iraqi and my children would be raised in Iraq; however, their higher education, in the tradition of my family, would be the best that I could provide for them, which would be abroad. Linda Turner and I had shared a joyously deep love relationship for several years. Our souls made love. Upon fully grasping the determination of my decision, we painfully parted and she married in 1967.

The galvanizing issue that solidified my political consciousness was Palestine.

The Palestinian struggle occupied a major part of my extra-curricular activities at both Michigan State University and the University of Michigan through the Arab Students Union. My Letters to the Editor would constantly appear in the students' newspaper. Commemorative events would be hosted in full political regalia. The Arab students held a yearly conference that was well attended by hundreds of Arab students from throughout the US. Our vociferous reaction to the Zionist distortion of the truth that permeated the US media (and still does to a very large extent), was the focus of all our concerns and activities. The political tempo was high and the pan-Arab feeling was throbbing and alive. My brother, Walid, who is more publicly palatable and diplomatic than I, was one of the main pillars of that pan-Arab organization in the early sixties. As a result of these events, heightened political considerations were augmented by the acquisition of effective organizational skills.

The defeat in the 1967 Arab-Israeli war was a major turning point in my life. The coupling between Israel's occupation of Palestine with American diplomatic and military support, fermented and sustained by American Zionists, was branded on my comprehension for life.

Driving along the highway to Ann Arbor, Michigan upon digesting the news of the military defeat, a long sustained scream, volcanically erupted from the bottom of my soul. Words no longer sufficed.

My actions in the Arab Student Union became bolder. Hearing of a United Jewish Appeal gathering intended to gather money for Israel, I attended it with Jean Littow, my life long and treasured friend. After listening to their initial speeches, and as they passed the envelopes around to be stuffed with money, I left my seat and walked to the center of the podium. Taking hold of the microphone, I appealed to the audience to consider the plight of the Palestinian refugees and urged them instead to donate that money to the Palestinians. As the initially shocked sponsors moved to block me from the audience, I walked to the edge of the podium so as not to loose eye contact with the astonished audience. Jean rushed outside to appeal to my cousin, Ramzi al-Saigh, who had come with us to that gathering, for assistance. He refused. Jean called the police and they promptly arrived. Seeing them approaching along the central isle with Jean, I jumped from the stage into their midst, clasped the arms of both as I sandwiched myself between them, and accompanied them out of auditorium, with Jean at my side. The Arab Student Society did get donations in the amount of $4,600 from that encounter which was then sent to Palestine. Several more phone calls threatening even my life were received; as such calls had already started to occur years earlier from my activism for Palestine at Michigan State University.

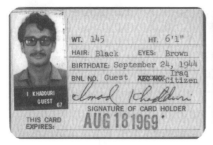

Identity card at Brookhaven National Laboratories, Upton, New York, 1967.

Tainted by political literature, I thought that the salvation for our condition lies in the economic reality. I dropped my PhD pursuit in nuclear physics that I had started earlier in 1967, at Brookhaven National Laboratory at Upton, Long Island, New York, and signed up for an Economics major at the University of Michigan in the fall of 1967. Capitalist economic principles quickly snuffed that line of action.

The Lebanese political activist, Hassan Cherif, arranged the contact for my joining the Palestine Liberation Organization, Fattah, in Jordan. Displaying all my belongings on the lawn of our abode, I sold everything that I owned and bought a flight to London and started to hitchhike to Beirut. An English gentleman offered me the opportunity to drive him to Paris. It was during the summer of the 1968 student upheaval. He had stacked up on petrol in the trunk of his car, since the French petrol stations were closed. He had dreaded driving on the right hand side. I more than gladly accepted his offer to drive his car, and he was a gentleman indeed from his dinning mannerism that punctuated our trip. Once in Paris, I wandered the streets watching bouts of stone throwing students who were attacking and then retreating in front of the police. At one point, and as I turned a corner, I found myself behind a police line. A policeman glanced behind him, noticed me, rushed to grab me by the scruff of the neck, and dumped me in one of their beating and bashing patrol cars. I pleaded my innocence to the police officer in my broken French, swinging my hotel keys, as he swung his baton for its usual trajectory. His superior stopped him just in time and asked me to show him my hands. They were clean, unlike the stone and cobble throwers' who had dirt and mud on theirs. I was set free.

Depositing a bunch of postcards addressed to my father with a lady friend of mine in Beirut, Lebanon, I headed to the guerrilla training camp of Fattah at al-Salt in Jordan. She would mail one postcard every week to my father, assuring him that I was on one of my usual summer trips.

Abu Salah was an able fighter. He supervised our training in guerrilla warfare in his intense and effective manner. There were about sixty of us. A few died after an Israeli air raid on our camp and a few during operations. Celebrating a visit of Abu Ammar, the head of Fattah, to our camp, Abu Salah held a race for him. He asked ten of his agilest to be fully laden with 50 kilograms of explosives and grenades, and asked them to race to the top of an upward slanting hill. My physical condition was such that I could literally feel my feet barely touching the ground as I flew up that hill, and won. Abu Ammar was impressed and chatted with me. He was surprised to learn of my background. The next day, Abu Salah called for me and delivered Abu Ammar's order that I should return and work for Fattah from Baghdad, as Abu Ammar had felt that I was too valuable, at that time, to remain a guerilla fighter. My objections did not resonate, and I sadly left for Baghdad.

I had missed Iraq so much that the blistering desert wind that scorched my hand extending over the window in the taxi taking us to Baghdad felt like waves of ice-cold water for their sensed familiarity. Once in Baghdad, I easily persuaded a taxi driver to take me home where I would secure the fare for him. My mother opened the door, looked at me from the distance as I stood by the outer gate and asked: "Who are you?" My shaven head and close to dark chocolate colored skin, from the rigorous training, eluded her. The taxi driver refused to accept his fare so as not to disturb the joy of my mother and father, who had emerged after her.

They had suffered deep stress, since my father had suspected something amiss after receiving many postcards from Lebanon. At the instigation of my father, my uncle, Khadduri Khadduri, had traveled to Syria a month earlier and contacted a friend of his in the Syrian Interior Ministry. His friend had retrieved transit records and confirmed that I had passed through Syria, from Lebanon, and had gone to Jordan.

My father, though greatly relieved at my return, was deeply hurt at my decision to drop science and the pursuit of a PhD. His main aim in life was "to build two castles; Walid and Imad with PhDs in their hands". Nevertheless, he most graciously left me to lick my wounds and feel my way.

After a few months of campaigning for Fattah in Baghdad and other Iraqi cities, by organizing and holding many inspiring Palestinian plays in theatres to garner their generated revenue, a serious disagreement developed regarding the handling of that revenue and I ceased my activities with the Palestine Liberation Organization.

CHAPTER FOUR:

Nuclear Reactor

The first step of a long journey

On a brisk autumn evening in 1968, Basil al-Qaisi, a dear friend from high school, sat down next to me while I was playing Backgammon in an open-air café meters away from the Tigris. He had heard that I had returned from the US, where I was studying physics since 1961, via a sojourn in Jordan. Sipping his tea, he dropped a suggestion that changed the course of my life. In his gentle, shyly provocative manner, he asked, "Why don't you join us at the Nuclear Research Centre? Our friends are already working there, Jafar Dhia Jafar, Nazar Al-Quraishi and others".

I was thoroughly taken aback. I was not aware that the Russians had built a two Megawatt research reactor at Tuwaitha, 20 kilometers east of Baghdad that went critical a year earlier during November 1967. I myself was at a decisive juncture, an indefinable point at the cross section of two lines; one of aspiration for an advanced education in nuclear physics on the one hand intersecting with fervent political convictions and commitments on the other hand. The proposed invitation warranted serious consideration.

After surveying the nascent research projects on inelastic neutron scattering augmented by state of the art scientific instruments and apparatus, operated with the youthful drive and vigor of several high school colleagues and assisted by a few International Atomic Energy Agency (IAEA) sponsored scientists, I decided to join the group of the eminent Iraqi physicist Jafar Dhia Jafar. Jafar had finished his nuclear physics PhD in record time from the University of Birmingham, England just a few years earlier. The total staff at the Nuclear Research Centre at Tuwaitha amounted to about 120 members who would have lunch together under the watchful and generous eyes of our Director, Ghazi Darwish, a respected and magnanimous chemist. Short meetings would ensue that hosted scientific lectures as well as frequent managerial planning sessions. The atmosphere was fragrant with enthusiasm, drive and high hopes.

After spending several months doing extensive research, I decided to forgo for the moment my political commitments and to continue my educational pursuit for a PhD, to the great relief of my honored father. My brother Walid was already finishing his thesis in International Relations at Georgetown University, Washington DC under the eminent scholar and historian, Majid Khadduri.

Hence, at the beginning of the summer in 1969, and after finishing at least one research paper[10], I took a leave of absence from work at the Nuclear Research Centre. Without any financial support or university acceptance, I doggedly headed back to the United States to attempt to rejuvenate my terminated PhD work in nuclear physics, that I had dropped earlier in 1967, at Brookhaven National Laboratory, New York. However, this would not be the case.

The flight made a stop over in London. I decided to break the journey and head for the University of Birmingham where my Lebanese friend, Mohammed Mikdashi, was studying. We had shared a living quarter during the last years at the University of Michigan, and he had followed his PhD supervisor's transfer to the University of Birmingham in order to finish his own thesis in mathematics. In the style of my travel at the time, I hitchhiked to Birmingham. Not knowing Mohammed's address, I headed toward the Student's

---

[10] *Experience with a Three-Crystal-Pair and Anti-Compton Spectrometer for (n,)* *Spectroscopy.* Jafar D. Jafar, Imad Y. Khadduri, et. al. Proc. of the International Symposium on Neutron Capture-ray Spectroscopy, Studvik-Sweden, 11-15 August 1969. IAEA, STI/PUB/235. Vienna 1969.

*Jafar Dhia Jafar, on the left, was the scientific head
of the nuclear weapons program, 1997.*

Union cafeteria, glanced at those assembled and spotted some Arab students. After the customary Arabic greetings and the invitation to sit down for some coffee, I asked about my friend. One of them did know him. Mohammed was living at a student-housing complex several kilometers away, but the student knew neither Mohammed's home location nor his phone number. As it was nearing five o'clock in the afternoon, and knowing Mohammed's habit of taking a stroll every evening, I got the directions for the bus that traveled by the housing complex. Before the fall of darkness and peering from the bus window as we passed by the students' housing, I spotted Mohammed's gait along the road, still bearded, deep in thought and with a pipe still slinging from his mouth.

Having gotten over the initial rush of meeting again after about two years, we turned to my plans. Mohammed skewed his head and shot that alarming glare at me when I mentioned that I intended to return to the United States to continue my education on a topic related to nuclear reactors. In his usual logical tirades, he pointed out that my return there would not go down easily with the pro-Israelis at the University of Michigan. As a student, both at Michigan State University and at the University of Michigan, I was deeply involved with the Arab Students organizations promoting pro-Palestinian activities. After the 1967 war, my anti-Israeli student activities climaxed. I had received several death threats. Mohammed was convinced that the knowledge of my sojourn in Jordan would seal my fate, or rather my life, should I return there.

"Why don't you stay here at the University of Birmingham?" Mohammed suggested. "They have recently opened a new Masters of Science course in Nuclear Reactor Technology, and it does have reputable professors". The next day, I went to visit the casual scholar Dr. Derek Beynon, a Lecturer in the Reactor Physics Group in the (then) Department of Physics and Astronomy at the University of Birmingham. I explained to him my previous educational experience and my interest in completing a PhD course. He enquired about my degrees. Having none on me, since I did not anticipate being with him, I recalled that I did carry in my coat pocket a letter of recommendation from a certain Jafar Dhia Jafar with whom I had worked for about a year. His eyebrows lifted, and still clutching on his pipe, he asked to see that letter. He read it, picked up the phone, called the university Registrar and instructed him to offer me a PhD acceptance but to wait for the required university degrees to come at a later time.

Beynon explained that Jafar had finished his PhD at the same Physics and Astronomy Department four years ago. He had left an honored impression with his PhD thesis on strong nuclear interactions accomplished in the minimum regulation time. Jafar's one page letter of recommendation was more than sufficient to secure for me acceptance in a PhD program at the University of Birmingham. I could not help but marvel at the glaring difference between the straightjacketed American requirements for acceptance to a PhD program and the trust based British educational approach.

A year later, Jafar, who was at that time the head of the Physics Department at the Nuclear Research Institute and a member of the top level Iraqi Atomic Energy Commission, paid a visit the Physics Department at the University of Birmingham. During the customary daily afternoon tea break at the Department's tearoom, I was sipping my tea with Jafar and the venerable Professor Burcham, the Head of the Physics and Astronomy Department. Professor Burcham had just put down his cup of tea when he turned seriously to Jafar and in all modesty, assured him that a position at the Physics and Astronomy Department would always be ready for him whenever he would decide to return and join, with Jafar sheepishly glancing an askance look at me with his soft smile as if imploring me to get him out of that delicate situation. Jafar returned to Baghdad, Iraq. Shortly afterwards, however, Jafar did leave Iraq to join an international research team of scientists embarking on a complicated nuclear physics experiment at CERN, the European Centre for Nuclear Research in Geneva, Switzerland.

Beynon assured me that no sort of financial assistance would be provided. With the University of Birmingham's acceptance in my hand, I hitchhiked back to the Iraqi Embassy in London seeking a scholarship. Mr. Barnes, the staid English gentleman in charge of the educational section at the Iraqi Embassy, politely responded in his most polished administrative posture that rarely had the Embassy granted a scholarship from London. In any case, I would have to apply for and get a scholarship from Baghdad. Undaunted, I telexed my father. The next day he went to visit Rashid al-Rifai, the Oil Minister, whom my father knew was a friend of Walid, my brother, during the sixties. (Walid was, until the mid sixties, in charge of the American branch of the Ba'ath party, with Rashid as a member Ba'athist.) The following day, Mr. Barnes received a telex instructing him that a Ministry of Oil scholarship had been granted to me pending a formal aptitude examination at the Embassy. Mr. Barnes took a look at the results of the exam administered to me, lifted his eyes in admiration over the rims of his glasses, and stoically stated that it was the best results he had seen for a long time. I would have to wait, however, for the formalities to be completed in a couple of months, but my student position at the University of Birmingham was secured. I left for the United States to try my luck there.

End of a dream

I was filled with apprehension upon landing in New York in the summer of 1969. It was not only a PhD that I was seeking. I needed to convince myself fully whether or not my impressions of the seven years that I had spent studying in the US and the total exposure to the "American Way of Life" that accompanied it, was or was not as negative as when I had left it two years earlier. I had faintly hoped that the lapse of the two intervening years would have afforded me some maturity and detachment from the youthful passionate distraught impressions that had so heavily accumulated upon me.

The events of the following two months in the US sealed my scruples. I had hitchhiked all the way from New York to Ann Arbor, Michigan and from there on across the US to Berkley, California, then headed north to Seattle, Washington and finally thumbed back east to Ann Arbor,

Michigan. It confirmed my decision of a few years earlier not to marry anyone but an Iraqi and to raise my children in the warm and generous atmosphere of the deeply rich Iraqi Arabic culture.

The Dean of the Physics Department at the University of Michigan, though welcoming my return and willingly offering me a seat, informed me, however, that all across the US financial assistance for graduate students at the end of the sixties was in tight straits. I would have to find my own financial support for the approximately five years of PhD studies. Nor would he be able to offer me a thesis in nuclear reactor physics, but instead in nuclear physics. Then he recalled that a physics colleague of his at the University of California at Berkeley had just recently informed him that he had a few remaining financially supported vacancies. He gave him a call, but unfortunately, he was unreachable, since we were at the start of the summer break. He advised me to get over there and present my credentials personally while he would try to get hold of his colleague to introduce me to him over the phone.

I left Ann Arbor with a heavy heart, for I found out that two of my close American student acquaintances had died; one had committed suicide and the other died from an overdose of drugs. I hit the road to Berkeley. I came very close to death twice during the three days and nights of continuous hitchhiking, once as a consequence of the Vietnam War, and the other due to my own fault.

Two men had picked me up in Ohio. They sat me between them in the front seat of their large pickup truck. They reeked of beer, cursing and swearing. They had just robbed a gas station earlier in the day, killed the owner, emptied the cash register and stuffed a couple cases of beer in the trunk of the car. They were on a killing spree as that was not their first killing. Driving fast, an unfortunate truck driver had noticed their erratic driving and saw them swigging the first cans of beer. They exchanged some dirty hand signals. The truck driver informed the police. The two were stopped by the police, but were allowed to continue on their way after proving they were genuine Vietnam War veterans and claiming to be freshly returning home from overseas to meet their families. Brandishing their guns, they were driving very fast intent on catching up with that truck driver in order to kill him as they were convinced that he had informed the police on them. Every second word they uttered was interspersed by "fuck". Suddenly, one of them asked me how much money I had on me, which prompted me to whip my wallet and show them the solitary five dollars bill in it. The rest of my money I would usually hide in my socks when I hitch-hiked. In order to further appease them, I started to imitate their manner of speech. I delicately maneuvered to ask them why were they doing this. Their stone faces reflected their deadpan response. They had killed so many in Vietnam, foe and civilians, that they could not, with their dead conscious, live any other way than to continue killing. They needed to keep pumping their Adrenalin rush and provide for their drinking habit. They had become killing machines. This aberration jives with recent quotes from American soldiers in Iraq: "Goddammit, I hate fuckin' Eye-raq and fuckin' Eye-raqis. I just wanna go home!"[11], and "We were told that once we entered Baghdad—and we won the war—they would send in other units to do the peacekeeping", one senior soldier said on condition of anonymity. "Those who did the killing should not be the ones keeping the peace. They need to send us home"[12]. The two drunken soldiers needed lots of therapeutic attention, a very scarce commodity for

---

[11] *"Occupation" is an ugly word"*, By Borzou Daragahi, The Iranian, June 11, 2003, http://www.iranian.com/Travelers/2003/June/Baghdad/

[12] *"Morale Reportedly Flagging in Iraq"*, The Guardian, May 28, 2003, by Chris Tomlinson. http://www.guardian.co.uk/worldlatest/story/0,1280,-2723308,00.html

Vietnam veterans at the time in the US, and probably still is now.

Feeling more comfortable with my presence, the two Vietnam veterans invited me to join them on their killing rampage promising me the bounty of several gas stations' cash registers. Feigning the urgent need to relieve myself and with the full support of their beer-laden bladders, they stopped at a rest area cursing the lost time in their pursuit of that hapless truck driver. They were headed south to the Mexican border. I was heading west to Berkeley, California. With great persuasion, I convinced them that we should, therefore, part our travel ways. Threatening me if I would ever reveal them to the police, and, with all the filthy words I could muster, swearing high and low that that would never happen, I painstakingly managed to extricate myself from them.

Several hours later, while sitting in another car some fifty kilometers down the road from that rest area, I saw the rolled over truck surrounded by police officers dragging the dead truck driver from his cabinet. I let the Americans take care of their dead, and their walking dead.

A day later, and in the early hours of morning along a Nevada highway, I was picked up by a cowboy. He had just finished a spate of cow branding and was generously paid for his work. He had bought a new car and was in fact heading to California which was written on the sign I held up for all cars to see. Since he was very fatigued, he invited me to drive while he slept. I did mention that I had been hitchhiking for two days and nights, but to no avail. He slept heavily next to me while I tried to stay awake along that pencil straight Nevada highway that does not bend an iota for kilometers on end.

The white painted stripes forcefully closed my eyelids that felt heavier than lead. Suddenly, reading the highway signs, I realized that I had been unwittingly sleeping while driving, with my hands firmly on the wheel on that never bending road. Alarmed, I decided to stop and sleep at the next rest area, which was about ten miles down the road, no matter what the cowboy said. The drag of dozing off is relentless and at such times uncontrollable. I slept at the wheel again and the road bent. Suddenly, I felt my balance swerving to the left. I opened my eyes and noticed immediately that I was quickly smashing several diamond shaped yellow signs that alerted drivers to the bending of the road. We were beyond the left hand gravel shoulder and heading into the ditch separating the highway. The car was starting to roll over into the ditch with the two right wheels off the ground already. I could see the pebbles of the ditch flying up from the two wheels that were still touching the ground. The immediate thought that seized me was that I would not want to die this way and hung onto the wheel with all my power. The cowboy jerked out of sleep and grabbed with all his might the bottom of the dashboard as he clung for his life. My life or death grab of the wheel managed to swerve the car out of the ditch and back on the road. The car stalled, straddled across the highway. I could see the lights of other cars heading right into the middle of ours, at full speed. I turned on the ignition and stomped on the gas peddle. The new car jumped like a gazelle to the right shoulder. Other cars whizzed by, their drivers not believing their eyes.

We got out of the car and surveyed the damage caused by the scratches of the yellow posting signs on the side of the car. "OK, partner, it is my turn to drive". I expressed my regret at the damage. He generously said not to worry as his next job would more than cover for it.

He drove us to a coffee rest area. Suddenly, the after shock hit both of us simultaneously as we sat on the counter. Our hands shook and the cups rattled and spilt their coffee. We were both uncontrollably trembling. When we finished cleaning the coffee mess and headed back to the car, we took a good look at it in the sunshine. The new steering wheel had fractured at both joints from my persistent life-or-death strength that I had applied to it against the force of gravity and other physics forces.

I finally reached the University of California and met the astute professor. He was indeed approached about my case but, unfortunately, the last seat was allocated during my travel to meet him. I headed north to the state of Washington to visit a lady friend, Gail Malizia, who was living on a boat in Seattle, while my heart was still pounding with love for Linda Turner, my love at the university, who has now ended living there.

Contacting Ann Arbor where I had left a forwarding address, I was informed that all the papers for the UK scholarship were in order and that I should head back to Birmingham University to properly enroll for the PhD course. I hitchhiked back to Ann Arbor to pick up my belongings. Along the way, I could only imagine in my mind the majestic beauty of the Rocky Mountains in Montana. It pained me to see their gutted and scarred slopes exposing their dirty entrails and crisscrossed by haphazard highways, a sad silent witness to the mindless exploitation of their resources by an environmentally hostile economic system.

On July 16, 1969, and after bidding farewell to my brother in Washington DC, I watched the first landing on the moon by Neil Armstrong, Michael Collins and Buzz Aldrin on television before embarking back to London.

By then, my base repulsive feelings and instinctive fear of the "American Way of Life" was indelibly stamped in my soul. I felt as if Pierre Vergniaud's prediction that the French Revolution, like Saturn, would devour its own children, would aptly apply to the beast in the United States of America.

Tragically, thirty-five years later, it would also devour my beloved Iraq.

Differing experience

My immediate supervisor at the University of Birmingham was the eminent Malcolm Scott, the Senior Lecturer and Supervisor of the Masters of Science course in Reactor Physics and Technology. Brushing aside as too esoteric the Masters in Physics that I so laboriously gained at the University of Michigan over a period of two years, Malcolm insisted that I take a second one-year Masters of Science degree in the specific field of Nuclear Reactor Technology. At the end of that one-year course, I would then be evaluated as to whether I would be able to continue for the PhD program. Four Iraqi students were enrolled for the Reactor Technology Masters course in 1969 at the University of Birmingham, Tariq al-Hamami, Abdallah Kendoush, Riyadh Yahya Zaki and I, all on generous government scholarships.

The adjustment to the British manner of teaching resulted in a marked increase in my studying hours. I did forewarn Malcolm Scott that I was fed up with taking exams and did not have the urgings to excel in them any more. At the end of that year, and upon the completion of our summer research project, Malcolm met with each one of us privately to inform us of his decision on where we stand. Tariq and Riyadh were to leave and find another university. They went to Glasgow. Abdallah had one of the highest grades in the examinations. He would stay, but under the burden of proof of his research capabilities. I, who had one of the lowest exam grades of the whole class, was to remain for the PhD program as my research project during the summer had apparently showed a good potential for continued advanced research.

After six months, Malcolm summoned Abdallah and told him to leave the Physics and Astronomy Department as it was deemed that he would not be able to rise to the standard of the required professional research capabilities required by the Department. Abdallah requested to be transferred to the Mechanical Engineering Department at Birmingham

University. Malcolm was adamant in his judgment. Abdallah excelled in passing exams by sheer memorization potential, yet his research capabilities would not satisfy the requirements of any Department at the University of Birmingham. Broken hearted, Abdallah joined Riyadh and Tariq at Glasgow.

The subject of my thesis was the measurements of the absolute fission rates per source neutron in an iron assembly using Solid State Track Detectors. This required the use of the 30 years old Nuffield cyclotron, which eventually closed down in about 1999 after 51 years of service. During my PhD research, it was kept running by the tender loving care of Fred and Ted.

There was still enough time to re-immerse myself in student politics, meaning the vehement defense of the Palestinian cause. That did irk other British students not sympathetic to the same cause, but was looked upon with interest by Malcolm Scott and Derek Beynon. To Malcolm's bewilderment, it also attracted several lady friends, among them Naomi Goldstein. At the end of one demonstration in London in front of the Israeli Airways El AL, we ended up in jail. Naomi, who was Jewish and an aspiring lawyer, rushed from Birmingham and managed to get me out of jail the next day.

In London, I was deeply involved, to no small measure, with Signe Collinson, the sister of Leslie whom I had befriended at Michigan State University in 1962 and whom I visited, hitchhiking, in Denmark in 1965, where she was by then living with her husband.

To London also came Jean Littow, my life long and treasured friend. She had so bravely stood by me at the University of Michigan after the war of 1967, in front of her fellow Jewish audience, while I tried to dissuade those assembled at a United Jewish Appeal meeting from donating money to Israel and tried to persuade them to send it instead to the Palestinians.

I was also graced with two "sisters", the Greek dentist Lefki Christidou and the Turkish physicist Yeter Goksu. Lefki was a bundle of sparkling nerves and joy, razor sharp comments and an engaging beautiful smile. Yeter was petite, chain smoking, with dreamy gazes that portended her intellectual insights. We bonded for life. On our honeymoon in 1976, my wife, Niran, and I first visited Lefki in Greece for two weeks. Lefki then accompanied us for another two weeks to Turkey to visit Yeter.

When I finally defended my thesis in December 1973, I had come to temper my repulsion for the heightened alienation of the West. I was indeed impressed by the deep and solid English traditions that shaded the English from the blistering heat of the American dollar, though rendering them comparably cold to my Arabic temperament.

Peaceful nuclear research

I rejoined the Nuclear Research Centre at Tuwaitha in the first week of January 1974. On the same day, Khalid Said, a fellow PhD physicist who had studied in the UK, had also started his work there and was immediately assigned to be the Head of the Nuclear Research Centre because of his Ba'athist status. I was never inducted as a full party member, but was in my high school days a "sympathizer". Muyasser al-Mallah, a fellow University of Michigan physicist, who by then was Head of the Iraqi Atomic Energy Commission, plainly told me of this stark fact. Had I had a similar party status, I would have probably been assigned instead of Khalid. Not that I looked forward to it. I was intent on doing research. Such a position entailed deep immersion in administrative work and routines for which I had little inclination, or anticipated to receive much satisfaction from it. Besides, I had severed my party connection back in 1962.

I joined Mansoor Ammar and Muqdam Ali in the Reactor Department and started to deploy my newly acquired scientific technique of Solid State Track Detectors in novel ways. These detectors were special plastic cellulose nitrate polymers that were affected by the passage of heavy charged particles, such as alpha particles, but they are not affected by gamma rays and beta particles. Upon etching in a base solution, the tracks of these alpha particles in the plastic would be enhanced and could be seen and counted under the microscope. Used with certain converters, neutrons may also be accounted for. I took pictures of small objects, such as bullets in order to look at the explosive powder within them, using neutron beams from the reactor[13].

While attending a scientific conference later that year, in 1974, and presenting my work on Neutron Radiography, I became aware of a method using these same cellulose nitrate films for the detection of uranium underneath the ground. The detection method depended on the disintegration of uranium, which would emit gaseous fissile products. These would filter through cracks in the earth layers and reach the surface, where they might disintegrate according to their half-life times and emit alpha particles. By placing inverted plastic cups with pieces of cellulose nitrate film attached to them, these gaseous products would be collected over a period of time, say a month, and when they would disintegrate, the emitted alpha particles would be close enough to the cellulose nitrate film to penetrate it and hence be detected and counted when etched. By placing the cups in a defined grid, the density of these alpha tracks would indicate, through a mathematical modeling of the grid points, the distribution contours of any uranium ores deep beneath the surface.

Upon my return from the conference, and on my own initiative, I immediately proposed to Khalid Said that I search for uranium in Iraq. He immediately concurred and assigned

---

[13] *A Neutron Radiography Facility on the IRT-2000 Reactor.* Imad Y. Khadduri
Nuclear Instruments and Methods, 147 (1977) pps115-118

Omran Mousa, the faithful and devoted driver, a vehicle, communication equipment, official papers and finance.

I first scoured the mountains in the northeast of Iraq, near the Iranian border. It was around the Kurdish village of Hero, still politically tense, but among the most beautiful natural scenery that I have ever witnessed. I would have 50 soldiers spread around in a circular formation, with me at the centre, fanning along with me as I planted the cups. The yellow uranium ore was even visible on the surface. What was most amazing were the littered and rusted drilling machines strewn over the mountainside. The local old Kurdish men and women spoke of the English lugging these heavy machines on mule backs over very rugged terrains during the thirties, and leaving them there as a witness to the importance of the abundant natural resources of that area.

Next, Omran and I headed south and spent several months in the Jil, an area in the barren desert on the Iraqi-Saudi border, and about 100 kilometers south of the infamous political prison specifically built in the middle of the

Prospecting for uranium near Hero, 1976.

desert, at the oasis of Nuqrat al-Salman (The hole of Salman). Siroor Mirza, the Head of the Geology Department at the Nuclear Research Institute at the time, accompanied us with his detailed maps during the first incursion to point to us the potentially suspected uranium ore laden areas in the middle of the desert.

Our last stop before venturing into the desert was the city of al-Samawa, about 200 kilometers to the north of Nuqrat al-Salman and straddling the vast southern expanse of sand that stretched way inside Saudi Arabia. There is a nearby small lake about 25 km west of Samawa, called Lake Sawa, whose water level is at least one meter above the surrounding flat ground and in the middle of the desert (strange enough, in Arabic the name Sawa is close to the word: level with). It is of interest because of its unusual physico-chemical properties. The water is salty. Upon evaporation in the heat of the desert, the waves licking the shore would deposit their salted crystals in accumulated layers. After many centuries, for the lake is mentioned as far back as two millenniums; the salty deposited rim of the lake had gradually built upward to a meter in height and still contained the raising water level. Very rare small tiny fish, not longer than five centimeters, survive in the lake. Their skeleton would show through their flesh. They were blind and would melt into fat upon heating. The lake is very deep. It is apparently fed by underground seepage from as far north as the Razazah Lake but has no surface outlet. The rumor has it, however, that the lake is connected with other similarly salt lakes through deep underground water streams. That fish species is only found in a similar lake in northern Russia.

While preparing for our venture in the desert at al-Samawa, we were naturally scouting for a guide to navigate the bewildering desert tracks. A policeman working at Nuqrat al-Salman police station presented himself and proclaimed his ability to take us there. Siroor hired him. Unwittingly, another good wisher had whispered in Omran's ears that this policeman was no Bedouin and that he had only wanted a free ride to get back to his post at Nuqrat al-Salman. Ominously, the whisperer warned that this policeman did not know the desert, contrary to his claim. Omran, for once, failed to relay that bit of information to us. Siroor telegraphed ahead to Nugrat al-Salman police station that we would be arriving by late afternoon.

After navigating us through the zigzag of earth trails that crisscrossed the desert as grooves on an old man's skin, the police "guide' finally put his head down in shame and claimed, "I am dizzy", the Iraqi slang for I am lost. He had no idea where we were. Our food and water supplies were meager. Wolves were around. We had heard in al-Samawa of the two Italian engineers who had slept over in that area a couple of weeks earlier. They were later recognized only by their boots along with some bones. Our "guide' lost his nerves and started crying. Siroor comforted us with the thought that he had in fact sent a telegraph to the police station in Nuqrat al-Salman and it would logically follow that they would miss our arrival and promptly inform the Iraqi Atomic Energy Commission back in Baghdad who would then send helicopters to search for us. In that anticipation, I drove the four-wheel drive to the top of a hill, and flashed the headlights in the direction of the sky every 15 minutes, in the faint hope that the search helicopters might be trying to locate us. The doors of the car were firmly locked against any prowling wolves.

Morning dawned without any sight or sound of a helicopter. Suddenly, a Bedouin on his camel approached us out of the blue. He professed that he was suspicious of some flashing light beams in the sky during the night and had been scouring around the area trying to find their source guessing from their projected traverse. It must be romantic watching the night sky in the middle of the desert every night. Thoroughly elated by his visit, perhaps we did not kiss him enough.

When we finally arrived to Nugrat al-Salman, the head of the police department was surprised to see us, especially with us tagging along his feeble constable. Siroor demanded, "What about the telegraph that we had sent you yesterday from Samawa informing you of our arrival?" The head of the police department was thoroughly taken back as he had not received it. Apparently, the policeman in charge of the telegraph machine has had a grudge against the "guide" that we had brought with us. Upon seeing the guide's name listed, he simply crumbled our telegraph and threw it away in disgust into the waste paper basket, where we found it. He had missed, in his blind antipathy to our "guide", the all too alarming phrase, "a team from the Iraqi Atomic Energy Commission". Both of them became really sorry because of their animosity to each other.

On the other hand, Abu Hamza, our experienced Bedouin guide, was formidable. One day, Omran and I had car problems and night befell us by the time we had fixed it. We clamored back in and drove in complete darkness attempting to reach our camp. Abu Hamza was sitting between me and the driver, out of respect, in order to allow me the luxury of the cool breeze by the window after a long hot day. After a short while, Abu Hamza could not take it any longer. He apologetically asked for the privileged permission to sit by the window. I obliged. He put his head outside the window, looking keenly at the ground passing beneath him under the floodlights of the car. As we moved, he gave instructions to Omran, to turn left, to keep going straight ahead, and to turn right. He was so familiar with the earth beneath him, the pebbles, the color of the ground, the small rock mounds as if he was looking at a detailed map. He finally exclaimed, "Stop". There was nothing ahead of us but total blackness with the lights of the car stretching ahead to nowhere. "Turn a bit to the left," he continued. Still there was nothing apparent. "A bit more and we will see a ruined mud hut," he finally pronounced. As the car turned, there was the mud hut 20 meters ahead of us.

Even more fascinating was the next morning. As we woke up and got ready to leave, Abu Hamza took a wide sweeping look and confidently stated, "This is Iraq", pointing to his feet, "and that there is Saudi Arabia" gesturing to a point five meters away. Thoroughly puzzled, I asked Abu Hamza how he figured that out. He grasped a shovel, walked a few steps and started to dig. At a depth of about 20 centimeters, a white powdery gypsum line appeared. "This is the demarcation line between Iraq and Saudi Arabia", he blithely explained. He was present as a young lad when the Wahabi Saudis drew this gypsum borderline in the desolate desert during the twenties, separating the two countries, and for hundreds of kilometers.

An unforgettable lesson in Arabic generosity graced me during that expedition. Omran and I were again lost in the desert. Scouring the emptiness around us, we noticed a small black dot in the distance. We headed for it to find a tent with a Bedouin, his wife, two children, a few sheep and a couple of camels, all in the middle of nowhere. We disembarked to his generous and warm greetings. We asked for directions. He insisted that we rest for a while in front of his tent, unfurling a rug for us to sit on. As we sat politely on the rug, he brought a water jar and started to sprinkle the dirt in front of us. Knowing fully well that water is more precious than gold in the desert, I shyly ventured to ask why this gesture. He stated, "To make the air more humid for you and to relieve you, my guests, from the heat".

After drinking his coffee, and getting directions on how to return to our camp, we tried to excuse ourselves. Not without eating dinner with him, he insisted. I tried to extricate ourselves with great difficulty from his offer, and asked Omran to give him what remained of our water supply. We started to get in our car. He ran away and before we had driven off, he had opened the back door and flopped a sheep in our trunk. "If you do not accept my dinner here at my tent, then take the sheep with you". We stayed for dinner.

We had to mark the locations of the measuring plastic cups, which were laid one kilometer apart in straight grid lines that stretched from ten to twenty kilometers. Omran and I resorted to building markers in the shape of small mounds of rocks on top of each buried cup, to the chagrin of multitudes of scorpions who rushed to unwelcome us from underneath the shade of the rock that they were suddenly deprived of. Our heavy boots must have died tens of times from their spiked poisonous injections.

As we dug the holes for placing the inverted cups at a depth of about thirty centimeters, we often came upon small seashells or the remnants of bigger ones. It is believed that in ancient times the sea did cover the whole southern desert all the way up to Najaf, which is about 150 kilometers south of Baghdad. To the south of Najaf, beyond the still evident few meters high rift, the area is still referred to by the locals as "the sea".

Omran and I then headed for the city of al-Qaim to the west, near the Syrian border. Nearby, we struck it rich. The results indicated heavy uranium concentrations near an area called Akashat. Apparently, it is a very rich phosphate deposit, and uranium is an assured by-product of such formations. A city arose in Akashat where a phosphate production plant was erected. One of the plant's buildings was for the extraction of uranium ore in the form of yellowcake. The extracted by-product would later be transported by rail north to our al-Jazeera site, near Mosul. That was the location of our processing plant which requires yellowcake as feed material in order to produce pure nuclear grade uranium dioxide, which in turn is chlorinated to produce uranium tetrachloride. This is the feed material for the Baghdatrons (a name derived from Calutron which is itself derived from the contraction of CALifornia University cyclotRON). These are the central pieces of huge magnetic separators that form the core of the Electromagnetic Isotope Separation (EMIS) process that was to be primarily employed by Iraq for the production of its weapon grade uranium 235.

Many months later, I was back at the Nuclear Research Centre with my findings[14]. A new challenge materialized almost immediately. The first Iraqi International Conference on the Peaceful Uses of Atomic Energy was to be held and I was one of its organizers. This event proved to be an important milestone for securing the services of a talented Egyptian scientist, and I will return to that shortly.

In the meantime, Khalid Said, who was then the Head of the Nuclear Research Centre, and knowing of my special relationship with Jafar, pleaded with me to approach him in order to persuade Jafar to return to Iraq. Jafar had by then joined an international team of over a hundred scientists on a nuclear physics project being implemented at CERN, Geneva in Switzerland. I complied and did write to Jafar. For reasons only known to Jafar himself, he did decide to quit his post at CERN and return with Phyllis, his English wife, to rejoin the Iraqi Atomic Energy Commission. There is, however, a recent bit of information from Robert Windrem that was reported on NBC NEWS on April 20, 2003, "Jafar had applied for a professorship at Imperial College early in 1975. Whatever his scientific virtues, however, he did not land the professorship. It was a fateful decision. Rebuffed by academe, Jafar decided to return with his family to the land of his birth".

In the meantime, I also got married. A mutual friend of

---

[14] *On the Use of Cellulose Nitrate Film in Uranium Exploration.* Imad Y. Khadduri, Iraqi Atomic Energy Commission, Nuclear Reactor Dept., Baghdad, Iraq. December 1979.

mine, the ebullient Maysoon Melek, introduced me to a petite, beautiful and dignified lady, Niran Shawket Jerjees. All three of us had lived on the same street in a suburb of Baghdad. Quite often, Niran and I ventured, despite tight social taboos and under the guise of darkness, into the Tigris River that was just at the end of our street. We would spend many memorable nights in my inexpensive Russian built motor powered dingy, landing on clean, fresh sand islands that would rise in the summer in the middle of the Tigris with nothing but a straw mattress and the moon. One night, as we ventured up the Tigris, our dingy strode upon a shallow submerged island, in front of the President's Residential Palace, with bellows of water churning upward from the stuck motor that was emitting a deafening roar. Niran kept her cool. I was most impressed.

We married on the spur of the moment, intentionally. Disregarding cumbersome and extravagant traditional wedding rituals that did not particularly appeal to me, we surprised our parents and close friends by announcing to them that we intended to wed the following day in the church. After the marriage ceremony, we offered some sweets and bid our guests a thank you and a farewell. They were welcome to visit us later at our new home. We had it furnished completely at a cost of 56 Iraqi Dinars, the equivalent of about $160. All furniture items, beds, couches, dinning table, closets and chairs were hand made in traditional manner from the stem of palm tree branches, an inexpensive item in Iraq.

When her side of the family visited us to offer their best wishes, her father, the dignified lawyer, Shawket Jerjees, leaned over and whispered in my ear, "You have married a very deep woman". It was poignant and true. In Arabic parlance, that would translate into, the more you polish a jewel, the brighter it shines.

We both decided that Niran was ready for higher education. We agreed not to have children until she got her Masters of Science degree in Computer Science. With the assistance of the late Abdul Wahab al-Kayali, a prominent and influential intellectual who was prematurely murdered a few years later, Niran got a two-year scholarship from the Iraqi government to finish her studies at Aston University in Birmingham, UK. She left in 1978 to the gracious welcome and hospitality of Malcolm Scott, my PhD supervisor who eased her settling down in Birmingham, while I stayed in Iraq. We rejoined two years later when I started my training on the newly purchased French research reactors at Saclay, France. Our daughter Yamama, the Arabic name of the bird that Noah set forth to bring him an olive branch indicating the end of the Great Flood, was born at the start of the Iraq-Iran war in September 1980. Yamama grew up with an astonishing resemblance to the calm beauty of my mother, as well as her incisive intelligence.

During 1976 and 1977, Iraq implemented its intentions of acquiring a nuclear electric power generating plant by assembling a scientific team to investigate the most feasible offer for a nuclear power plant. I was part of the team that met the sup-

Mr. Ito, in the centre, and the Mitsubishi team, 1977.

pliers' delegations when they visited the Nuclear Research Centre and held wide ranging and detailed negotiations with them. The Iraqi team visited several potential nuclear power plants in Japan (Mitsubishi), Sweden (ASEA Atom) and West Germany (Kraftwerk Union AG). We had made great strides in our negotiations with Mitsubishi at their headquarters in Tokyo, Japan. We were nearing the end of it, when the venerable Mr. Ito, the head of the Japanese delegation, excused himself after listening to a hushed aid whispering in his ears. He went out for five minutes, and returned to declare the end of the negotiations. Westinghouse, the American Company that supplied the nuclear fuel for most Western and Japanese nuclear power stations, had just called to refuse any supply of nuclear fuel for an Iraqi nuclear power plant. We returned to Baghdad with a lesson well learned. If we are going to get a nuclear power plant, it will have to be Russian. Little did I know then that that recommendation of ours would engulf me for several years in the next decade.

Dabbling with critical mass

The first Iraqi International Conference on the Peaceful Uses of Atomic Energy was held in Baghdad in the spring of 1975, under the coordination of Hamid Auda, an affable nuclear biologist. Being in charge of the reactor technology sector, I oversaw the evaluation of the submitted papers and allotted the time for them. My attention was directed to an Egyptian nuclear reactor scientist who had submitted about ten papers. Being suspicious at first of the relatively large number of works submitted by one person, and anticipating trivial scientific solutions, I took the time, nevertheless, to carefully read his submitted papers. His extensive work and the depth of his analysis were most impressive. I asked him to try to combine all topics into one lecture, offering him a full hour to deliver it. Even Malcolm Scott, my PhD advisor, who was delivering our joint work from the University of

Birmingham, and Yeter Goksu, my life long sister, were allotted fifteen minutes each.

Yehya al-Meshad, the most eminent Egyptian physicist, displayed his scientific prowess effortlessly. An excellent lecturer, he presented the complex scientific material with clarity and ease. He had also intended to come to the conference to apply for a teaching position during his sabbatical leave from Alexandria University in Egypt, and had approached the University of Technology in Baghdad. We immediately made sure the following day that his application was duly accepted. An enduring warm friendship quickly developed with Yehya.

Malcolm had suggested that we start a one-year Reactor Technology Master of Science course based on the material that he had developed for his course at Birmingham University. He would be willing to accept, simply upon my recommendation, any graduate student of this course for a PhD program at Birmingham University. His trust was humbling and he did stand by his promise in the coming years.

Coordinating the dynamics of the course with the University of Baghdad, we launched the Master of Science course in Nuclear Reactor Technology in the following fall. Two students enrolled. They would be completely under our guidance at the Nuclear Research Institute, but their degrees would be conferred by the Physics Department at the University of Baghdad. Our main pillar for the course was Yehya al-Meshad for the theoretical part. I oversaw the experimental part, based on Malcolm Scott's materials and manuals. The two students, who would journey several times a week to the University of Technology to attend Yehya's lectures, reflected to me the valuable manner with which he lectured the difficult material. I bought them fifty tape cassettes and requested that they tape all his lectures for posterity sake. I still treasure these tapes, though I had to leave them behind me in Baghdad.

Seeking a valuable scientific experience from Yehya's presence in Baghdad, I arranged for a car to bring him twice or three times a week to the Nuclear Research Centre, and to relieve the effort of our two students to travel to him. Once at the Centre, I also engaged him in developing a computer program, or code, to calculate the burn-up of the reactor's nuclear fuel instead of depending on the simplified hand calculated formulas that were left to us by the Russians[15].

Yehya's breadth of knowledge and experience in nuclear reactor technology was abundant and overflowing. Our code and calculations opened up the possibility of calculating critical mass, the correct density at which a highly enriched uranium 235 sphere would undergo a self-sustaining chain reaction; this could become a reactor, if controlled, and an atomic bomb, if uncontrolled. Bouncing off the idea with Khalid Said and Jafar Dhia Jafar met with encouragement. In addition, we started dabbling with different implosion scenarios that would start with a smaller spherical sphere of uranium but would increase its density to a critical value. In a classical nuclear bomb, an outer sphere of explosive material is detonated at a precise moment to create an inward directed shock wave, focused on the spherical fissionable sphere. This very powerful shock wave would compress, or implode, the sphere into a smaller size, and hence higher density. A thermal neutron from an accompanying neutron source would be readily absorbed by a fissionable nucleus causing it to fission and releasing energy as the mass of the resulting fission products is less than the mass of the original nucleus. This small amount of mass difference is converted to energy according to Einstein's famous equation $E=mc^2$. When the imploded sphere reaches a critical mass, it is assured that out of the 2-3 neutrons being emitted as a result of the fission of this one nucleus, exactly one of them

---

[15] *CORELOAD: A Computer Code for Calculating the Evolution of the Operation History of the IRT-2000 Reactor.* Imad Y. Khadduri, Yehya al-Meshad Iraqi Atomic Energy Commission, Nuclear Reactor Dept., Baghdad, Iraq. Report No. NR-11, June 1976.

would cause another nucleus to fission. This fissioning process is rapidly repeated, in a very short time, in a self-sustained chain reaction. The bomb explodes, releasing intense amounts of energy and radioactive fission products.

We gauged our results with actual experimental results of highly enriched assemblies that were carried out during the forties for the Manhattan Project's atomic bomb, such as Godiva and Jezebel for bare spheres and Popsy and Topsy for spheres surrounded by neutron reflectors. These experimental results were already published in the open scientific literature. Our calculations matched the experimental results[16].

Upon the completion of Yehya's two-year contract with the University of Technology in 1977, I could not miss the chance of having him completely working with us full time. My proposal on this matter was accepted on the proviso that I would find out why he would want to stay in Iraq. One afternoon, as I visited him in his house to drink tea and look over some scientific journals, he pointed to the pictures of some researchers that were posted near their published articles. "These were my students", he said. They were all by then in the United States and prospering scientifically. I grasped the opportunity. "Why would you not want to go to the United States when you have such good contacts there that would assure you of a research position?" His teenage daughter had just then entered the room with another round of small glasses of tea. He thanked her politely and turned to me, "It is because of her and her sister. I have been hearing too many sad stories from these students of mine in the United States of how they have lost control of their children there, especially their daughters. Iraq is a Moslem country and I feel very comfortable in raising my children here". Yehya was to be hired by the Iraqi Atomic Energy Commission.

---

[16] *The Use of Multigroup Transport Method for Criticality Calculations of Some Fast Spherical Assemblies.* Imad Y. Khadduri, Yehya al-Meshad. Iraqi Atomic Energy Commission, Nuclear Reactor Dept., Baghdad, Iraq. Report No. NR-14, June 1978.

Excited by the prospect of Yehya's eminent arrival at the Nuclear Research Center, I scurried around collecting and preparing all the required papers and documents that were to be signed. Having piled them up on Hamid al-Bayati's desk, our legal advisor, I left Yehya with him awaiting just the final signature of Abdul Razzak al-Hashimi, the Head of the Iraqi Atomic Energy Commission at the time.

Abdul Razak is a calamity that seeks self-destruction. We nick named him, Chouqi, and that is how I will continue to refer to him. Chouqqa, in Iraqi Arabic slang, is when you aim a marble at a collection of resting marbles, smashing into them and dispersing them uselessly all around you. Chouqi is an Arabic language inflection take on that fitting description of him.

Hamid paged me urgently as I worked in the reactor building. Chouqi had decreased Yehya's salary from 450 Iraqi Dinars to 425 Iraqi Dinars. A government minister salary at that time was about 400 Iraqi Dinars. Yehya was thoroughly insulted, had picked up his leather handbag and was heading for his car. I literary ran from the reactor building to the gate, about 200 meters away. On the way, I bumped into Khalid Said. Still running, I explained to Khalid, in halting breath, what had transpired and urged him to pay a visit to Chouqi immediately. I caught Yehya as he was closing his car's door, with a few security officials running behind me not knowing what in the world was happening.

Yehya was rightly insulted. "It is not the money, but the indignation of that offer," he explained in trembling voice. As I calmed him down, Khalid drove to the parking lot in his car with Chouqi's signature on the full amount.

Having mastered the tools for calculating the burn-up rate of the nuclear fuel in the reactor due to Yehya's diligence, Jafar beckoned that he and I should jointly carry out a detailed calculation on the possible production of weapons grade fissionable plutonium 239 from the operation of the Russian reactor's fuel. Plutonium 239 constitutes the core of another type of atomic bomb. This was to be carried out without Yehya's participation. It was a long shot. Plutonium 238 is also produced in the nuclear fuel rods and it is a very good absorbent of neutrons, a "poison" that would impede a sustained chain reaction of plutonium 239. At a certain low percentage of plutonium 238 "contamination", a mixture of plutonium 238 and 239 would be rendered useless for an atomic bomb core. Hence, it was crucial that the fuel rods in the reactor be allowed to burn a limited time only to preserve the required ratio of plutonium 239 to plutonium 238 for an atomic bomb. With our low power research reactor, it would have taken decades to obtain the required amount of nuclear weapon grade plutonium. The relevance of the work, however, was the knowledge of the required calculations[17].

Independently, I was working on a very complex experiment to absolutely measure the burn-up of the nuclear fuel cells, employing a large stainless steel tank and the transport of highly radioactive "burned" fuel cells, those that have been used to run the reactor for some time, in five tons lead containers. The fuel cell would be inserted inside the tank, manipulated and rotated in a precisely controlled motion for two to three days. The measurement of the radiation level at certain distance intervals would indicate the amount of uranium 235 that had fissioned, or burned, and the remaining amount of unburned fissile uranium 235 in that fuel cell. The research project could only be completed with advanced calculations made possible through our training on the OSIRAK French reactor.

In 1974, a top level Iraqi government delegation, lead by Saddam Hussain, visited Paris. That visit was dubbed as the Masgoof visit. A planeload of Iraqi specialty fish, along with

---

[17] *The Possible Production of Pu^{239} from the IRT-5000 Reactor.* Jafar D. Jafar, Imad Y. Khadduri. Iraqi Atomic Energy Commission, Nuclear Reactor Dept.,Baghdad, Iraq. October, 1978, published in 1983.

their chefs and the special firewood required for the feast, were flown to Paris to exhibit Iraqi hospitality. The fish were cooked in full regalia, as they were held vertically flat on sticks in front of a bonfire, a delicacy in Iraq called Masgoof. The team discussed the purchase of two research nuclear reactors.

In 1976, a high level Iraqi delegation visited Paris to seal the negotiations. The delegation was headed by Chouqi and consisted of Jafar Dhia Jafar, Hussain al-Sharastani (the chemist who was later jailed for about 11 years and escaped from Iraq at the end of the 1991 war) and Humam Abdul Khaliq (the rising star of the nuclear program). The French labeled the two reactors after the Egyptian gods OSIRIS and ISIS, after the nomenclature adopted at the Saclay Nuclear Research Institute near Paris where the designs for these reactors were to be prepared. OSIRIS was a relatively large forty Megawatts research reactor and ISIS (the sister and the wife of OSIRIS in Egyptian mythology) was a one Megawatt mock-up of OSIRIS employed to test the functionality of the experimental setups before bringing them in for proper irradiation at OSIRIS. The name OSIRAK was bestowed upon the whole project according to French accommodating sensibilities. Iraq simply called them TAMMUZ 1 and TAMMUZ 2. All major revolutions in modern Iraq occurred in the very hot month of July, which is Tammuz in Arabic.

In anticipation of the expansion of the Nuclear Research Centre to include the two new research reactors, attention was turned to the proper defense of the Tuwaitha site from a military attack. The Soviet influenced military thinking at the time called for the erection of a three-story high earth mound that would completely surround the enclosed area, save for two 100 meters long tunnels. If these two tunnels ever needed to be blocked, the enclosed area would be completely protected, even from a devastating flood. Tremendous effort went into the construction of this seven or eight kilometers long by 25 meters high dirt mound, which was interlaced at 5-meter intervals with layers of welded steel meshes for stability.

The rising Atomic Energy star Humam summoned me. "What would happen if a bomb hit and breached the containment of the Russian built reactor? How far would the radioactively contaminated cloud reach? Would Baghdad itself be affected?" I set forth to do my calculations. In order to verify the results of my theoretical model and account for various wind speeds that would affect the dispersion of the emitted cloud and its radioactive fallout, I asked for, and got immediate permission, to access the wind tunnel at the Military University in al-Rasheed military camp nearby. We built a scaled down model of the whole site, with all of the buildings and the completed mound. The model was placed inside the wind tunnel, smoke would be controllably emitted from the reactor's "chimney", and various controlled air speeds would be applied across the model to watch the dispersion of the smoke of the "radioactive cloud". We took video shots of the whole series of experiments, at various wind speeds. The results were close to our theoretical calculations. Most importantly, I found that if the mound was covered by trees, the absorption of the radioactive isotopes by the leaves and greenery was dramatically increased, hence lessening the level of the radioactive fallout. Humam ordered the installation of a controlled water irrigation system over the surface of the whole mound, which was later planted with trees.

In the mid-nineties, an Iraqi physicist, Khidir Hamza, managed to escape from Iraq and seek the tutelage of the CIA. At the end of 1999, he published a book titled "Saddam's Bomb Maker". It is worth mentioning that at no point in time did Khidir Hamza get involved in any research work related to the nuclear bomb or the effects of a radioactive accident when we dabbled with such research. His mere solace was in the never-ending three-body problem, which was very irrelevant to such a weapons program. There is not a single documented scientific report of any work by Khidir Hamza relating to critical mass or a nuclear bomb in the arch. ` of the Nuclear Research Centre for that period. His book w. uld give an otherwise impression.

Ma' di Shukur Ghali Obeidi, a solid state materials scientist, was assigned to organize the scientific and engineering team that was to be trained at Saclay on the operation of the two reactors and on the six experimental rigs that were the prime reason for buying them. I was assigned two of the most complex of the six rigs, IreneAkis[18] and MarineAkis[19], for the testing of the characteristics of the nuclear reactor fuel under extreme operationa! conditions. Mahdi was later assigned to head the centrifugal enrichment process team in the eighties. At the end of June 2003, he led the Americans to some hidden documents and components of the centrifugal process under a rosebush in his garden.

Five graduating students from our ongoing Nuclear Reactor Technology course joined me at Saclay, France. About 60 scientists, engineers and technicians were dispatched in early 1980 to the research center at Saclay to learn French in an accelerated language course followed by one year training on the operation of the two reactors and the six experimental rigs. Basil al-Saati was in charge of the whole team while we were in France. Unfortunately, Basil was not up to the task. He could not make decisions on his own, but was simply a conduit for paper trails that would be quickly transported in front of him with his typical rubber stamp signature theme, "For your review and awaiting your instructions". This extremely weak managerial stance caused

---

[18] *Neutronic Calculations for IRENEAKIS. (Work performed in Saclay, France)* Imad Y. Khadduri, et.al. Iraqi Atomic Energy Commission, Nuclear Reactor Dept., Baghdad, Iraq. Report No. PH-RP-P02-81-1, January 1981.

[19] *Neutronic Calculations for MARINAKIS. (Work performed in Saclay, France).* Imad Y. Khadduri, et.al. Iraqi Atomic Energy Commission, Nuclear Reactor Dept., Baghdad, Iraq. Report No. PH-RP-P03-80-2, December 1980.

The group of graduate students from the MSc course in
Reactor Technology at Saclay, France, 1980.

an ominous rift in the team.

Serious disagreements were fermenting between the training staff and the French. Foremost among them was the sudden switching by the French of the type of the nuclear fuel that would be used in the two reactors. Instead of the 80% enriched cylindrical elements, specified in the purchase contract, we were going to get *fait accompli* an 18% "caramel" type fuel, named after the much-honored French chocolate bits. The French designed the low enriched "caramel" solely for this project for Iraq in order to be completely sure that Iraq would not be able to use the original highly enriched fuel for nuclear weapon use. Though they had signed a contract to deliver the highly enriched fuel in the contract a few years earlier, they immediately started to design the new type of fuel and surprised us by its offering in 1980. We were supposed to be testing that new type of fuel in IreneAkis and MarineAkis.

Iraq, understandably, was thoroughly irked at this switch in the terms of the contract. They sent Yehya al-Meshad to renegotiate the terms of the contract. On June 13, 1980, the Israeli Intelligence Agency, the Mossad, smashed Yehya's head with a copper rod as he entered his hotel room in Paris. The only witness, a French woman, was "mysteriously" run over by a car and killed a few days later. With immense grief, the whole Iraqi training contingent accompanied his casket to the airport. My wife and I then attended a Requiem Mass by Gabriel Faure in a Latin Quarter church in Paris. Never had I cried so much.

Strong disagreements also developed among those in charge of the training team. My vociferous criticism of the poor qualifications of some of the Ba'ath party members that were part of the training team soon reached Baghdad through the swift conduit of Basil's rubber stamp, "awaiting your instructions". Riyadh Yehya Zaki, the head of the Technical Team, who had to leave for Glasgow upon finishing his Masters degree at Birmingham University, stood in conspiratorial support in fear of the retribution of the Ba'athists in his own entourage. We were at the middle of our scientific training period. It was then that Chouqi had visited Paris while I was on a mission to Baghdad. He gathered the whole team and held a kangaroo court martial of my claims without me being present to be able to defend myself. "How many of you think that Imad Khadduri is wrong?" he asked the group. Most of the 40 Ba'ath party members, to the delight of those at the centre of this controversy, lifted their hands. "How many of you think that Imad Khadduri is right in his criticisms?" he continued. A few party members and the remaining independents lifted their hands.

Chouqi returned to Baghdad while I was still there but chose not to confront me with this fiasco, which was true to form in his usual uncourageous manner of dealing with events. Upon my return to Saclay from Baghdad, before the Christmas of 1980, I was informed of the kangaroo court theatrics. A few days later, a directive arrived from Chouqi ordering the termination of my training and ordering me to return to Baghdad immediately. The dilapidating news, which was detrimental to the success of the training on two of the most important experimental rigs, arrived in the morning. I was informed of it by Basil, who feigned regret with a dismissive face, before attending an important meeting with the French team headed by M. Genthon. I still attended the meeting and defended the Iraqi team's positions and claims effectively and persuasively *vis-à-vis* the French. When we left the meeting, the noble Basil Mahmoud al-Sammarai, the head of the Engineering Team, leaned over and asked me awkwardly whether I knew of the news of my recall. I nodded that I did. "And you still defended us so bravely in front of the French just now?" asked Basil with shock.

Basil al-Saati and a few of his loyal party members accompanied my wife, my three months old daughter Yamama and myself to the airport, at very close range and all the way to the doors of the Iraqi Airways airplane, lest we deviated unto another flight.

In the first week of January 1981, I returned to the Nuclear Research Centre. I immediately went to the center's expansive library, found a small unoccupied archive room, settled in with an Arabic coffee machine, Havana cigars and Samuel Glasston's book "The Effects of Nuclear Weapons", and commenced translating it into Arabic. Without paying the least attention to formal procedures, I did not sign any official paper signaling my return to any department or office in the Nuclear Research Centre. I would simply go to work every morning, head for my sanctuary at the library, translate the book and entertain the brave few souls with Arabic coffee when they dared to visit me. This situation was most unusual and unorthodox. Khalid Said and others went to Chouqi urging him to summon me to clear the air. Chouqi refused: "I will not encounter Imad in order to avoid what might be unpleasant consequences". In Arabic, it does sound more ominous (Likai atajannab ma la yahmud aqbah). I continued my self-imposed vigil until the genuine start of the Iraqi nuclear weapons program, which took place on September 3, 1981.

CHAPTER FIVE:
Nuclear Bomb

Make the bomb

Events, unknown to me at the time in early 1981, were already in motion behind the scenes concerning the fumbling goal of obtaining a nuclear bomb. The bright chemist, Hussain al-Shahrastani, was put in Jail in December 1979. We were already aware that Hussain was a devout Moslem. He would, in the middle of work at his laboratory in the reactor building, stop his work and spread a computer paper listing on the floor when prayer time arrived and kneeled in the direction of Mecca to pray. What we heard, or rather were allowed to hear, was that an operative of the al-Da'awa party, a fundamentalist and orthodox Islamic party that was deemed hostile to the government, had claimed that he had heard Hussain express disapproval of the widespread arrests of Shi'ites taking place at the time and that he was bringing back with him, from his scientific visits to France, al-Da'awa political instructions hidden in toothpaste tubes. It was well known to all of us at the time that the regime had zero tolerance for all political enemies. It was absolute folly to engage in political activities that aimed to topple the regime. It was tantamount to playing Russian roulette.

On December 4, 1979, Humam Abdul Khaliq, the emerging leader of the Iraqi Atomic Energy Commission and Khalid Said, the head of the Nuclear Research Centre, paid Hussain a visit at his office and engaged him till the arrival of several security officers. Jafar Dhia Jafar, who was by then an Advisor to the Iraqi Atomic Energy Commission, had accompanied Humam and Khalid, but was unaware of the purpose of the visit. The officers confronted Hussain and whisked him to prison.

The next day, Jafar paid a visit to Hussain's home to comfort his Canadian wife and their children and to see if there was anything he could do to help to put them at ease, as far as possible under the circumstances, in the hope that Hussain would be released soon. Upon ringing the door bell, two security men answered the door with their drawn pistols. It was standard practice to set up a trap in a detained person's home in an attempt to pick up any of his collaborators. Jafar and his driver Omran (my companion driver in the desert) were ushered into the house. Hussain's family was not at home. The security men called their headquarters by phone. They were apparently instructed to release Jafar and Omran. It was obvious that the house was thoroughly searched as papers and books were strewn all over the floor. Jafar decided to wait for the family to return. After about half an hour they did return and Jafar tried to calm them down and told them not to worry as this was a 'routine' investigation. Jafar stayed with the family until the two security men had left the house.

Hussain al-Sharastani was tortured and put in solitary confinement. Several months after his arrest, Barazan, Saddam's half brother and Chouqi paid Hussain a visit to his jail cell and offered him a reprieve and rewards if he would cooperate in a nuclear weapons program. It is claimed that Barazan asked Hussain to cooperate on the extraction of uranium from Iraqi phosphates and that Hussain replied that he would try his best[20]. It is also claimed that Barazan had warned Hussain, "Whoever is not willing to serve his country does not deserve to be alive" and Hussain's response was, "I agree with you it is a person's duty to serve

---

[20] *"Profile of Dr. Hussain Shahristani"*, by Eric Goldstein, Huqoqalinasan.org. http://www.mafqud.org/en/partners/hio/goldstein.htm

his country, but what you are asking me to do is not serving my country."[21]

Hussain remained in jail for the following ten years. Before the invasion of Kuwait in August 1990, Hussain's solitary confinement ended and he was allowed to mingle with other prisoners. He took that opportunity to arrange for his escape. In the confusion of the aftermath of the 1991 war, Hussain did manage to escape on February 13, 1991 and to take his family with him to Iran.

After Hussain's arrest, Jafar had appealed to Chouqi in his defense. Typically, Chouqi rushed over to Saddam accusing Jafar with intentions that Chouqi is well versed in fabricating. Saddam ordered the house arrest of Jafar in January 1980. Only many years later did I know that the house in which Jafar was confined in was only a few streets from my home in Baghdad. He lived there for twenty months with the constant company of an ever-present Intelligence officer. Though I was very intimate with Jafar, we never did discuss what had transpired during that imposed confinement. Upon my return from Paris in January 1981, and learning of his predicament, I started to visit his worried mother quite often to console her, as she herself confided a lot in my father, her medical doctor. I tried to provide some comfort with whatever scraps of information I could gather about his unknown whereabouts. Those were unsavory times. Humam Abdul Khaliq called me to his office. "If you do not stop visiting Jafar's mother, they will fry onions on your ass", he threatened menacingly. "They" referred to the Intelligence officers. Onions need a lot of heat to be cooked. Sustained hitting with a stick on the butt was Humam's insinuation for generating the heat required to cook the onion. I refused to obey the injunction and continued to visit Jafar's mother.

---

[21] "Hussein's Ex-Scientists Say Nuclear Bid Stymied", Boston Globe newspaper, Farah Stockman, Globe Staff. Date: 02/09/2003 Page: A1 Section: National/Foreign.

When Jafar was finally released, I was told that his venerable mother had told him, "You have a true brother in Imad".

Israel belligerently bombed and destroyed Tammuz 1 and Tammuz 2 on the evening of June 7, 1981[22]. The bombing occurred in late afternoon and after most of the staff had returned home. We heard the blasts and ran to the rooftops. We could see the cloud plumes even tens of kilometers away. We sadly watched the unchallenged Israeli warplanes streaking west in the setting sun. When I arrived for work the next morning, the security guard, fully aware of my standoff with Chouqi, refused my entry to the site and returned me home. Meeting with work colleagues later that afternoon, anger and defiance were chokingly building in our throats.

Israel had defended its belligerent attack as forestalling Iraq's attempt to use these reactors for the production of nuclear weapons. Two possibilities were postulated.

The first concerned the utilization of the fuel itself to acquire the weapon grade fissile material. The French had already made sure that the uranium fuel was so lowly enriched and intriguely designed as to be inappropriate for such use, even if diverted and intentionally enriched, a process not easily concealed under strict French and IAEA inspections of the fuel inventory. Furthermore, the provided fresh fuel amounts itself that was to be supplied to Iraq would not be sufficient for a full-scale weapons program even if such a process were initiated.

The second contention, based on Israel's own production of plutonium for their more than 200 nuclear bombs during three decades, was that Tammuz 1, the 40 MW reactor, would be used to produce plutonium. The possibility of such an undertaking by Iraq is delusional. The tight refueling schedule for such an endeavor, which is required to pre-

---

[22] Statement by the Government of Israel on the Bombing of the Iraqi Nuclear Facility near Baghdad, June 8, 1981.
http://www.mfa.gov.il/mfa/go.asp?MFAH0i5s0

vent "poisonous" plutonium 238 from developing, would be impossible to hide from the French scientists who would have been collaborating with us for years and the IAEA inspectors. Had we even diabolically thought of kicking both out and running the reactor ourselves for such a purpose, the limited fresh fuel that was allowed for us would have aborted any such attempt at the outset. Neither would the unique design of the reactor core for the "Caramel" fuel allow for fuel designs specific for plutonium production.

The only logical inducement for Israel to bomb these reactors would have been to prevent Iraq from obtaining scientific and technological nuclear expertise, but not nuclear weapons. In that, they apparently did not expect the gut Iraqi reaction. Get the nuclear bomb covertly and in spite of Israel.

Saddam took the political decision to initiate a full-fledged weapons program immediately afterwards. Jafar had to be resurrected from his internment. It took some time to convince Jafar, who was still by then confined and blistering from that experience, to consider, plan and accept the terms of his release and the scope of the work to be started. I believe that he wrote, while still interned, several technical reports on the matter to Saddam to that effect. When all of that had come to pass, he was issued his mandate and released.

Jafar arrived at the Nuclear Research Centre on the morning of September 3, 1981 signaling the start of the nuclear weapons program. One of his conditions for his return to work was the ouster of Chouqi. As Jafar entered, a car sped Chouqi away, never to set foot again at the Nuclear Research Centre.

Walking through the gates of the Nuclear Research Centre, Jafar asked for a few colleagues, and enquired about my whereabouts. I left my library sanctuary for good.

Basil al-Qaisy, Munqith al-Qaisy, Munqith al-Bakir, Zuhair al-Chalabi, Nabil Karnik, Imad Ilyia and a few others each sat down in a humble office waiting for a private meeting with Jafar. Assignments were handed down to each one in turn. After their meeting with Jafar, the two highly qualified technicians, the electrician Nabil Karnik and the superb welder Imad Ilyia, confided their doubts and scruples to me of what was expected from them. "Is he really serious?" they asked. "Can we ever design and build what he is asking us?" they continued, smilingly hiding their anxiety. Indeed, we were on the first steps of a determined full-scale nuclear weapons program.

Jafar asked me to set up solid procedures for documenting the scientific and technical reports that would be forthcoming from this effort as well as the procedures for the covert purchase of the required materials for the program in coordination with Natiq Butti, the highly experienced Head of the Purchasing Department at the Nuclear Research Centre. The liaison between Natiq and I was Ahmed al-Rihaimi, a devout stickler to overt rules and guidelines, but an unaccommodating obstacle to the required secrecy and covertness. With much pain, we were making our initial purchases of sensitive detectors and materials.

The procedures for documentation considered foremost the integrity and quality of the presented work. Jafar would rely on my scientific evaluation of the report's content before submitting it to him for a final review along with a recommended distribution list for his final approval. Special paper was used for publishing the final report, as well as its limited number of copies, that would indicate any unauthorized photocopying of any page of the report. Strict auditing procedures were laid out to make a physical inventory of all distributed copies every year. Three separate locations, two outside the Nuclear Research Center, were assigned as repositories each with a complete set of the reports.

At first, the organization that Jafar was setting up for the nuclear weapons program was assigned the title of Department 3000, Research and Development. Department

1000 was the office of the Deputy Chairman of the Iraqi Atomic Energy Commission under Humam Abdul Khaliq (Saddam Hussain was the Chairman, then). Department 2000 was International Relations headed at the time by the physicist Rahim Kittel. He was later appointed as the Iraqi ambassador to Austria, close to the International Atomic Energy Agency (IAEA) in Vienna. Department 4000 was Administration under Dhafir Selbi. Department 5000 was Projects and Department 6000 was the Nuclear Research Institute under Khalid Said. All Departments, except for Department 3000, were still carrying on with peaceful nuclear research under the watchful eyes of the IAEA.

Settling slowly into my new responsibilities in the nascent program, Jafar requested an urgent mission. He had gathered a list of important scientific articles and reports that were desperately needed. We also needed to purchase a couple of lasers to investigate the feasibility of enriching uranium using a novel technique. Accompanied by a polished young Intelligence officer, we attended a scientific conference in San Diego, California in 1982 on the use of Solid State Track Detectors, the materials that I had been using for research during the seventies. I immediately set out to search for assistance to obtain the required articles and reports. Left on one of the tables at the conference was a business card of a retired librarian who had access to the budding Internet and could search for such documents. I rented a car and drove to her home. Several days later, I revisited her and collected most of the desired articles for about $200 in cash, shook her hand and parted. I do not recall that she asked where I was from; neither did she question the nature of the gathered information.

I then dispatched my watchful companion to New York to wait for me as I went forth to purchase the lasers from Florida, having arranged the whole deal by phone from San Diego in California. I made a short stop at the airport of

Miami, Florida where I met an Indian technical sales representative. He handed me a briefcase containing the two lasers, and I in turn handed him a briefcase containing $30,000 in cash. The entire transaction lasted about an hour.

Back in Iraq, Niran's aunt woke me up in the wee hours of the morning in November 1981 announcing the arrival of Tammam, our son. He was named after the famous Christian Arab poet, Abu Tammam, who belonged to the Christian Iraqi tribe that lived in Iraq before the arrival of Islam. I would take Tammam along with me whenever I would visit my Arab tribesmen friends in Koufa in the South, or Sharqat in the North. Thus, he witnessed many instances of their renowned hospitality and tasted their delicious food.

Making solid progress in my purchasing activities and the documentation procedures, I slowly migrated to the small planning group that was working directly with Jafar on the nuclear weapons program. That led to a slight blunder that terminated, for a while, my work with Jafar. Part of the planning was the assignment of scientists and engineers to attend relevant and worthwhile conferences and symposiums abroad in order to glean the state of the art knowledge of the technologies and research in the fields that would support our program.

Hassan Cherif, my close Lebanese friend from our study days in the United States and my main conduit to the Palestine Liberation Organization that led to my sojourn in Jordan, had found a job at ESCWA, the United Nations Economic and Social Commission for Western Asia. He was stationed in Baghdad. He called to inform me about an important electronics conference that ESCWA was going to host in Kuwait. I nominated our foremost electrical engineer, Basil al-Qaisi. Unaware, Intelligence officials were monitoring Hassan's phone calls. They were extremely upset that I did not ask them for permission or immediately report talking to this "foreigner". I am not sure that I was even aware of the need of such required procedures at that time, especially with a close and trusted friend like Hassan. Intelligence would have none of it. They assumed that ignorance of their own secret regulations was no excuse for a breach. To them, that incident posed a most serious security risk. They ordered to have me kicked out immediately, and not only from Jafar's group, but from the Iraqi Atomic Energy Commission all together.

One morning, in 1983, and unexpectedly, a sudden order was delivered to me by a solemn Intelligence officer to drop my pen and leave my desk immediately. He ordered me to report to the Reactor Department, outside the scope of the nuclear weapons program. Ten years later, I finally became aware of what had transpired behind the scenes.

Jafar Dhia Jafar and Dhafir Selbi, my high school mate, who was then the Head of Administration at the Nuclear Research Centre, had vehemently opposed the Intelligence decision to exile me out of the Iraqi Atomic Energy Commission and to the University of Baghdad. Serious heated arguments had ensued. The compromise was that I would stay at the Nuclear Research Institute on the condition that I would rejoin the Reactor Department. Adnan Jerjees, the dedicated and dependable Head of the Reactor Department, was surprised at my arrival and at even further loss as to what to do with me. He left me to decide what role to play. I decided to walk the grounds of the Nuclear Research Centre for hours, days and weeks on end, alone or with a few brave souls who again supported me in my rebellious solitude. Humam Abdul Khaliq, Head of the Iraqi Atomic Energy by then, called for me for a reprimand, yet again, for this second obstinate gesture. Recalling my rebuff to his demand that I stop visiting Jafar's mother, he did not promise me barbequed onions this time. His concern was that my challenging stance would dampen the morale of others. He smoothly encouraged me to settle on some work. He probably was also keeping the Intelligence officers at bay, as best as he could. I reluctantly chose to do a re-evaluation of the safety aspects of the Russian research reactor in the light of its intended upgrade from 2 MW to 5 MW, which was being undertaken at the time. Within a few months, another more challenging opportunity soon materialized that pulled me out of this deep hole. At the end of 1983, I was transferred to the nuclear electric power plant project, which was put on a higher priority level, under the direction of Khalid Said.

I first met Khalid Said at an annual Arab students' conference in Fort Collins, Colorado in 1964. He, along with a dozen Ba'athists, was ousted from the Lumumba University in Moscow for reasons I never fathomed. The whole group was hauled to the United States to finish their studies. I ran again into Khalid when I used to hold get togethers for Iraqi student friends on many weekends in my flat in Birmingham, UK where I would cook Iraqi food, sweets and serve home made beer. Khalid was finishing his PhD program in Solid State Physics at a university close by. We returned to Baghdad with our degrees at the same time and joined the Nuclear research Centre on the same day. He was immediately elevated to be the Head of the Centre. Unlike me, he was a Ba'athist. Khalid Said died in a hail of bullets after he failed to stop fast enough at an American checkpoint in Baghdad on May 3, 2003. I would sincerely hope that arrogance and repulsion did not overwhelm the required sense of caution.

I started working with Khalid on the nuclear electric power plant project early in 1984. The International Atomic Energy Agency had established very detailed steps and guidelines for the various stages of selecting a site, the construction, the startup and the operation of a nuclear electric power plant. Each step required the implementation of a Quality Assurance program, which consisted of well-defined and documented procedures on how to implement each of these stages. My first assignment was to head the Quality Assurance Department at the Iraqi nuclear electric power plant project[23] for the site selection stage. This required several training visits on the concepts of Quality Assurance to Vienna, the headquarters of the International Atomic Energy Agency. It was very instructive, and it upgraded my awareness of the professionalism required for such an enterprise; a point not missed when I later rejoined Jafar's group in 1987.

The lengthy site selection process was estimated to take from five to six years. A lot of data had to be gathered locally; meteorological, seismological, hydrological, environmental, flood history, electrical, etc. I was in charge of contacting the various Iraqi government agencies to negotiate with them the contracts for carrying out these studies and surveillances. Many of them had not yet heard of Quality Assurance procedures. However, their data had to be gathered, calculated and verified in accordance with these procedures. That meant the promotion of these concepts in the various Iraqi government organizations, not a simple feat while dealing with civil servants accustomed to following the path of least resistance in performing their work.

I was settling in this line of work. Nofa, our youngest daughter, was born in June 1987. The name is pure Bedouin. One of its meanings is a tall woman, a graceful lady standing on a mound. There are other meanings for Nofa in the Arab lexicon. Nofa, even as a child, exhibited a remarkable sense of stability in difficult situations, a mirror of her mother. However, Nofa was measurably more extroverted, not in an aggressive manner but rather in a self-assured demeanor. When she was born, I instructed the nurse to name her

---

[23] *The Site Selection Quality Assurance Program for the Iraqi Nuclear Power Plant.* Imad Khadduri, Mohammad J. Abbas, Iraqi Atomic Energy Commission, Nuclear Power Reactor Project, Baghdad, Iraq. Report No. SA-IR-04-001, January 1987.

Umaima, another beautiful Arabic name, which does rhyme with her siblings Yamama and Tammam in Arabic. Apparently, Niran summoned the nurse as soon as I had left and crossed out Umaima and put Nofa instead. She explained later that she was looking forward to a boy whom she intended to call Nawfal, and Nofa is closest to that for a girl.

Things started to shift quickly. In the winter of 1987, I attended a high-level meeting chaired by Humam Abdul Khaliq, the Head of the Iraqi Atomic Energy Commission, in which he outlined the Commission's priorities. I was not aware then of the great upheaval that had recently transpired behind the scenes. As I left that meeting, my lasting surprised impression of it was the realization that the nuclear electric power plant project was no longer a priority and was instead destined to become a mere façade for the IAEA to focus on and follow, while the real nuclear weapons program would be surging forward with a new vigor.

What had actually transpired at the time was a crucial turning point in the Iraqi nuclear weapons program.

Petrochemical 3: PC3

To the extent of my knowledge, Khidir Hamza, the self acclaimed "Bomb Maker", had either failed in his assignment under Jafar's program to make solid headway with the gaseous diffusion enrichment process or had simply coveted Jafar's position as head of the program. I tend to lean to the former, as Hamza was certainly not qualified to lead any program. He was a loner, only adept at working on his theoretical three-body problem for more than two decades. He did not have the charisma or the courage to lead a team. His distaste of any experimental scientific work provided a focal point for many humorous puns.

In either of the above cases, Hamza had written a lengthy report to Saddam Hussein, at the beginning of 1987, accusing Jafar of procrastination and wasting the vast resources allocated to him. Saddam was furious and demanded an explanation from his top scientists.

It was at this critical juncture that Dhafir Selbi, who was until then the head of the Administration Department at the Nuclear Research Centre, was requested to join the top management team of the nuclear weapons program. He brought fresh new ideas on how to move forward. Initially, he suggested that the top management team should take a week off to the Habbaniya Lake resort to deliberate and reconsider their progress. After coming up to speed from Jafar, Humam and the others on how the work was being carried on in Department 3000, he proposed a breakthrough brainchild of his to restructure the mode of work relationships. Having the material resources and the human resources, he emphasized that it is the network of interaction that makes it possible to best utilize these resources. He proposed the creation of "zumras", an Arabic metaphor for "teams" that would be comprised of engineers and scientists, delegated by their various scientific and engineering departments, to tackle specific design proposals. Collectively, the zumra would work through and materialize the designs through collective interactive thought of all concerned scientific and engineering activities. This was in radical contrast to the previous mode of work where the design was put forth by one department and it would go by back and forth between the different disciplines who would merely append their thoughts to it individually, and not interactively.

The resulting restructuring at the end of their deliberations entailed the formation of the following groups:

Group 1: The centrifugal enrichment process, which was assigned to Mahdi Shukur Ghali Obeidi.
However, several months later, Hussain Kamel, Saddam's trusted son-in-law, took direct responsibility for that group and shrouded it completely under his own sphere and direction.

Group 2: The PIG and TIG enrichment processes (soon to be substituted by the Electromagnetic Isotope Separation (EMIS) enrichment process under the convincing critique of Dhafir Selbi) was assigned to Jafar Dhia Jafar.

Group 3: The "administrative support" group that would enlighten Jafar's administrative chores was assigned to Dhafir Selbi. This group was responsible for covert purchasing, the provision of scientific and engineering information, the documentation of the scientific reports, the mechanical and electrical manufacturing activities and in a later stage the supervision of their design activities. I was incorporated into that group in September of that year, 1987.

Group 4: Khidir Hamza was asked to drop the diffusion process, much to his delight, and was assigned to gather a team for the design of the nuclear bomb. However, he was soon kicked out after a few months and the nuclear weapon design group was assigned instead to Khalid Said.

The Petrochemical 3 (PC3) project was officially announced in the summer of 1987, replacing Department 3000.

A huge drab building, the former headquarters of the Iraqi Labor Union, adjacent to Hussain Kamel's headquarters at the Military Industrialization Corporation, was assigned to Hamza for his weapon design group. Within a few months of efforts to prepare the building for occupancy, Hamza availed himself to three of its air conditioning units and hauled them to his house. The security nabbed him. He was immediately stripped of all privileges and shipped back to the Physics Department at the Nuclear Research Centre to rejoin his three-body problem that was eagerly waiting for him.

During 1987–1989, I saw Khidir Hamza many times at the Tuwaitha Centre, descending from the bus along with the rest of the staff, going to his office, bow headed and a non-entity, instead of his previously customary arrival in a government limousine, with a bloated chest. He immersed himself in preparing a report on the American Strategic Defense Initiative gleaning what he could from open scientific literature available at the library. In order to make some more money, after being cut off from the well-heeled benefits that he had just lost, he desperately sought to obtain a permission to be sent to Poland in order to procure a dense plasma focus machine claiming that it could be used as a cover to get fast electronics for an implosion weapon. He had already planned to retire after that trip.

The total staff of "Saddam's Bomb Maker" in 1988 and 1989 amounted to one sole lady, a physics major who graduated in 1987. She was transferred from under his management even before he returned from Poland. The reason she was offered was his impeding retirement. Yet, in an interview with Hamza published in the Washington Post[24], he referred to Colin Powell's report to the Security Council on February 5, 2003, in which he claimed that in 1995, "as a result of another defector", the United States discovered Hussein's "crash program", initiated after the invasion of Kuwait in August 1990, to complete a crude nuclear weapon. "He was referring to me," Hamza boasts. Hamza, having fully retired from the Iraqi Atomic Energy Commission by that time, was a mere support lecturer at al-Mansoor University College in Baghdad, a far cry from his "bomb making" fiction.

In 1987, with Khidir Hamza kicked out of the role of the head of the weapon design team, Khalid Said was inducted to fulfill that role and had to relinquish the nuclear power plant project. The soft spoken, cigarette munching perfectionist Atta al-Rawi took over that job and became my boss at the nuclear electric power plant project.

That was when, in the spring 1987, I had attended that

---

[24] "*The smoking gun*", by Richard Leiby, Washington Post Staff Writer, February 6, 2003; Page C01.

high-level meeting to discern our new goals, only to read between the lines that the nuclear electric power plant project had dropped by the sideline.

In October 1987, and in order to gain a firm control of the nuclear weapons program, Saddam appointed Hussain Kamel, who was already the Head of the Military Industrialization Corporation to be in charge of Groups 2, 3, and 4. In addition, Hussain Kamel took a direct and separate leadership of Group 1 that was completely distanced from Groups 2, 3 and 4. Group 1 was to work on the centrifuge enrichment process under the continued direction of Mahdi Shukur Ghali Obeidi who surfaced in early May 2003. The activities of these four groups would be made completely invisible from the International Atomic Energy Agency.

In January 1989, PC3 was established within the Ministry of Industry and Military Industrialization (MIMI) under Hussain Kamil and included the whole of the Iraqi national nuclear program (enrichment and weapons). Petrochemical 1 and Petrochemical 2 were established large-scale refinery projects undertaken by MIMI during the eighties.

In contrast to Khidir Hamza's false claims, Jafar Dhia Jafar, Humam Abdul Khaliq and Dhafir Selbi, were, in my opinion, the true dynamic prime movers of the nuclear weapons program.

The ever-low profile Dhafir was yet another one of my high school mates, along with Basil al-Qaisi and Nazar al-Quraishi, whose paths crossed at the Iraqi Atomic Energy Commission. After graduating as an engineer from Baghdad University, he had joined the British run Iraqi Petroleum Company at the oil rich fields of Kirkuk. The disciplined and meticulous work habits of the British engineers, coupled with Dhafir's keen sense of management, initiative and foresight propelled him into a leadership role. He was transferred to the Iraqi Atomic Energy Commission in the late seventies. He was assigned, of all things, to be the Head of Administration. We numbered about 1500 at the time. Dhafir set in motion the laying of one of the most effective administrative infrastructures that I have ever come across in my entire work career, be it transport, inventory warehouses, work procedures, salary disbursement, or finance and purchasing. His methodical approach, coupled with his warm and compassionate human openness to colleagues and fellow employees, entrenched him in the hearts of all who encountered him. It is extremely rare for any administrator in such a position of responsibility, with so many people to deal with, which swelled to about 7000 by the end of the eighties, not to leave one disenchanted employee who would bitterly criticize him for a fault. Yet, this was indeed the case with Dhafir. I would humbly ascribe to him the major role for catapulting the nuclear weapons program into high gear, starting in 1987.

Dhafir was the overall seer of implementing our covert purchasing efforts. When the volume exceeded the stringent and inflexible stances of Natiq Butti, head of the purchasing department at the Nuclear Research Centre and his stubborn colleague Ahmad al-Rihaimi, who was the focal point between that department and PC3, Dhafir took a surprising decision. In his superb estimation of people's potential that was sorely missing from many around him, Dhafir chose Adil Fiadh, a junior nuclear reactor technical operator who had graduated from Germany, to head a new purchasing department solely dedicated to the covert purchases of PC3. It was an astonishing decision.

I recall an incident involving Adil Fiadh in the late seventies. We were once carefully inspecting a bulging tumor in the stainless steel casing of a neutron beam channel in the reactor hall with my newly purchased high radiation tolerant camera, when Humam irritantly turned to me and demanded "who in the hell is that nut head shouting at the top of his voice on the reactor floor above?" I paged Adil with my walkie-talkie instructing him to keep his shrill voice down and to calm down. He was freaking at the prospect of a break in the corroded bulge that would lead to the leakage of the radioactive water that cooled the reactor core. We were standing in front of that channel. At that time, Adil was not outstanding by any means, save for running an antique shop in his spare time.

Yet Dhafir spotted potentially untapped talent in Adil, to the consternation and ridicule of Natiq and Ahmad. The nascent purchasing department that Adil started with a few assistants burgeoned within several years into nearly ten separate fully functional covert fronts with a staff of fifty. He himself had several passports. He displayed immense propensity for tight administration, strict procedures, open ears and quick firm decisions.

Most characteristic of the devotion, integrity and discipline of the work at that time was the utter lack of, or news of, any improper commission taking or kickbacks in executing all sorts of financial transactions and contracts, from the highest authority, through Adil and down to his staff, even though billions of dollars were spent in a matter of few years. Adil was an important part of the success of the nuclear weapons program until its demise in 1991.

Adil Fiadh was murdered in his ranch outside Baghdad in 1994. Hundreds of colleagues respectfully attended the three days of mourning ceremonies at his home. Rumors abounded as to who committed the crime, ranging from Intelligence agents, to shady deals after the war due to the deteriorating economic situation, to bandits in the neighborhood of his farm. Khidir Hamza, for some odd reason, took the tragedy as a threat against himself. In his book, he cited Adil's murder as another prime reason for his own escape from Iraq at the end of 1994.

Khalid Said faced a serious dilemma. Jafar had gathered the entire first-rate scientists, especially the physicists and the chemists, under his wings. Those that were left still working at the Nuclear Research Centre were either below that caliber or specialized in fields that were not required by Jafar.

Within a few weeks of the assignment of his new role, Khalid called me over to visit him at the Nuclear Research Centre. He proposed that I would take over the Physics Department that he was intending to set up for the calculations of the nuclear bomb design, recalling clearly my early theoretical work with the late Yehya al-Meshad and with Jafar on this matter.

However, working with Khalid for three years during the nuclear power plant project did not prove to be palatable for me. As a result, I declined his offer to his great dismay and despite his persistent persuasion. My explanation to him was simple, "I can't work with you because whatever you ask me to do, by writing it in the morning in pencil, you would erase it later in the afternoon".

In the meantime, even though Dhafir was assigned the task of offloading the administrative load off Jafar, he was still part of the top management and bore the responsibility for the success of the project as whole. He delved into the research and the techniques called PIG and TIG that were the pet favorites of Jafar for the enrichment process for the previous six years. He came to the conclusion that it was indeed not a promising line of work to follow. Dhafir was an avid reader. He went to the library and read the literature. He consulted other scientists and engineers in the program. He reached a conviction that it was more feasible for us to switch over to the tried and proven Electromagnetic Isotope Separation (EMIS) method that employed huge magnets, referred to as Calutrons. This enrichment process was implemented during the Second World War in the Manhattan Project to produce the first American atomic bombs that were dropped on Hiroshima and Nagasaki in Japan.

After several heated and intense arguments, Dhafir managed to convince the top nuclear weapons management team of his newly arrived at conclusions. A decision was

Salam Toma, on my right, a most trusted companion and confidant, 1992.

made to go for the Electromagnetic Isotope Separation enrichment process and the Baghdatron, and as fast as possible. Dhafir had to prove that his choice of the enrichment process was the right one.

Hearing of my rebuff to Khalid's offer to lead the Physics Department at the newly formed Group 4, Dhafir called for me the following day. He went into the above scenario in great detail, uncustomary told to an "outsider", as I was still officially working with the nuclear power plant project, as well as being scorned at by the Intelligence Agency. I was so amazed by the details of Dhafir's exposition that I did venture to ask him, "Have the Intelligence changed their minds about my 'security risk'?" Dhafir brushed aside my query dismissively and got down to the basics, as he is so famous for. "Jafar's scientists are not doing their abc of scientific research," he explained. "They are tiring a bit after six years and are not properly researching published articles on their new assignment. I want you to flood them with proper scientific and engineering information. I also want you to take hold again of the documentation procedure. The scientific quality of some of our reports that I have seen should have been thoroughly reviewed and reworked before being approved and distributed". He assigned me one person at first, Khawla, a beautiful and sharp lady, with haunting almond black eyes. I soon added Salam Toma, my most trusted companion and confidant.

To ease my transition back into the nuclear weapons program and to avoid an unduly disruption of the nuclear power plant project's momentum, Dhafir suggested that I might split my time half and half between his assignment and the nuclear power plant project. Atta al-Rawi, in his perfectionist glory, utterly refused to share me. I bade him farewell on the spot and joined Dhafir. I headed Dhafir's Activity 3W, in Group 3, labeled Information and Documentation.

Information

The next day after accepting Dhafir's work offer, and after securing an office for Khawla and myself, I took a stroll in the expansive library of the Iraqi Atomic Energy keeping in mind Dhafir's detailed exposition and target. I noticed a long row of multi-shelved bookcases. It was the complete set of the United States' Nuclear Science Abstracts (NSA) from 1947 until its demise in 1976 when the National Technical Information Services (NTIS) morphed into the Department of Energy (DOE) in the US. The first few years of this series were given to us as part of the gift that was offered to Iraq in 1956 under the "Atoms for Peace" program promoted by President Eisenhower. They were part of a complete library of published literature on Atomic Energy at the time. The library of the Iraqi Atomic Energy had subsequently subscribed to the NSA and covered all of the intervening years, until the end of that series in 1976.

That precious atomic library of 1956 included another

gift: a small research reactor. It happened to be near its destination at the Iraqi seaport of Basra in the Arab Gulf in July 1958 when the 14th of July revolution took place in Iraq that toppled the Hashemite monarchy. The Americans could not anymore deliver this delicate cargo to a revolutionary government. They solved their dilemma by offering it as a gift to their freshly reinstalled Shah of Iran. The CIA had backed the toppling of the nationalist leader, Mossadaq, who had managed to rid Iran of the Shah but had dared to think of nationalizing the oil industry. That research reactor was, I believe, installed at the University of Tehran.

Finding the NSA treasure trove, I quickly went through the yearly indexes. I picked the keywords that covered what I thought was required for the EMIS project and the atomic bomb; words such as critical mass, Manhattan Project, Calutron, critical assemblies and so on. By that time, Salam Toma had joined my group. I handed the list to my two associates and asked them to go through the yearly indexes of the whole set, covering a period of about thirty years. Two weeks later, they brought me back a list of citations that filled more than fifty pages. The next task was to identify which of these citations were actually present in our abundant library. Ninety six percent of them were already there. The library's card index had also indicated that among the listed titles was the presence of about 30 items of the actual published set of more than 50 reports, books and microcards of the first American nuclear weapon produced under the Manhattan Project and published under the National Nuclear Energy Series. The actual list that I found (and still hold) on the content of the National Energy Series was probably printed by our library staff in the sixties. However, the catalogue card index did not indicate where they resided in the library alleys and reading rooms. It took me several days of searching for the keys of forgotten attics and storage rooms in the library building. In one of them, I found a box that was probably not opened since the sixties laying under a centimeter of dust. In it were the Manhattan Project books and reports. Further sleuth digging on my part revealed the total complete official list of the published books and reports in National Nuclear Energy Series that I had barely managed to copy from a microfiche. I immediately confiscated and locked the original books and reports.

Employing a few extra staff, several copies were made of the entire collection of books and reports. Classifying the available scientific literature by subject matter, a list was disseminated widely to our scientists and engineers, each according to his field of specialization.

Upon further investigation, two problems arose.

First, there were references to more than 160 patents submitted by the Project's scientists and engineers themselves that pertained directly to the Manhattan Project and the design and operation of the Calutron. These were in fact the actual designs, processes and descriptions of the vital components of the EMIS. They were not part of the Atomic Energy Library gift. We had to get them.

Siroor Mirza, our ever well dressed, with a flower pinned to his jacket sleeve, Scientific Attaché to the Iraqi Embassy in Vienna was ready. Siroor, a geologist, had joined me ten years earlier during my uranium prospecting trip to the desert in the south of Iraq. During that trip, we had shared the unsavory memory of the misguided guide that nearly cost us our lives.

In coordination with Siroor, I would send him a list of about twenty or thirty references for scientific articles, papers and patents. Dispersed in each list would be several Manhattan Project patents that could, at that time, be obtained from the World Intellectual Property Organization in Geneva, the container of all the patents of the world. I soon had collected all 164 patents pertaining to EMIS, classified them into categories, and distributed copies of them to

The complete published list of the books, reports and microcards of the Manhattan Project under the "National Nuclear Energy series", 1961.

(Translation)
Head of the Commission, Sir:
I attach herewith the results of the
patents' survey with gratitude to
your interest and follow-up in
obtaining them in full.
For your perusal, please.
With gratitude.

Imad Khadduri
21 July 1988

God bless your efforts, Imad.

Signature of Humam Abdul Khaliq
21 July
Seal of the Head of the Atomic
Energy Commission

A thank you note and (facing page) *a sample of the 164 Manhattan Project patents classified according to topic that were widely distributed and used in the Baghdatron design and operation, 1988.*

the scientists and engineers concerned. The Head of the Atomic Energy commission, Humam Abdul Khaliq, extended his thanks. It probably cost us no more than $100.

Second, some of the books were on microcard, a predecessor of the microfiche and the microfilm. This method of document imaging was used in the late forties and early fifties. Page images were miniaturized and printed on a postcard size glossy paper, just like a postcard, with about 20-40 miniaturized pages per card. I also found the original microcard reader that came with the library in 1956, under even thicker dust layers. Light had to be reflected from the surface of the microcard and the image had to be sufficiently magnified to be read. Notably, there was no way to print the output, but one would just barely be able to read the contents, as the image quality was very poor. A very cru-

1 - CALUTRON :	PATENT NO.
- THE COMPREHENSIVE PATENT ON CALUTRONS BY ERNEST LAWRENCE.	2,709,222
- CALUTRONS.	2,725,478
- RECTANGULAR CORE OPENINGS, CASTINGS OF LONG SHALLOW MAGNET TRAYS.	2,727,190
- CALUTRON ASSEMBLING AND DISASSEMBLING APPARATUS.	2,871,361
	2,871,362
	2,871,363
- MOUNTING AND SUPPORT OF SOURCE AND RECIEVER TO ENABLE QUICK AND ACCURATE ALIGNEMENT AND EASY ACCESS.	2,871,364
- HAND TRUCK FOR CALUTRON HANDLING.	2,874,860
- DISASSEMBLY AND DECONTAMINATION APPARATUS FOR CALUTRONS.	3,143,119
- CALUTRON WITH MEANS FOR REDUCING LOW FREQUENCY RADIO FREQUENCY SIGNALS IN ION BEAM.	3,260,844

2 - PLURALITY OF ION SOURCES :	
- CALUTRON WITH PLURALITY OF ION SOURCES AND COLLECTORS.	2,714,664
- PLURALITY OF TANKS AND MAGNETIC FIELD STRUCTURES.	2,721,272
- PLURALITY OF INDEPENDANTLY REGULATED ION SOURCES AND FILAMENTS.	2,733,347
- MULTIPLE ION BEAM TYPE.	2,754,423
- OPERATE PREDETERMINED PORTION OF SYSTEM FOR SEVERAL CALUTRONS.	2,847,576
- DUAL HEATED ION SOURCE HAVING ARC SHIFTING MEANS.	2,882,409
- DUAL ION SOURCES WITH LINER.	2,890,340

3 - ION SOURCE :	
- SUPPORT AND INSULATING ARRANGEMENT TO RELIEVE STRESSES.	2,714,165
- CLEANING OF SLIT OF SOURCE FACE.	2,714,665
- SIMPLE AND COMPACT ION SOURCE.	2,715,683
- ION SOURCE WITH TWO INSULATION PLATES, APPLY VARIOUS POTENTIALS TO THEM TO SHAPE AND AFFECT SURFACE OF PLASMA.	2,882,409
- ION SOURCE WITH REGULATION OF CATHODE, ANODE AND ION CHAMBER VOLTAGE.	2,733,340
- MECHANISM CAPABLE OF ADJUSTING ION SOURCE THROUGH SEVERAL PLANES	2,737,590
- SEVERAL CHARGE CHAMBERS CONNECTED TO SINGLE ARC CHAMBER	2,817,763
- IMPROVED ION SOURCE, GOOD HEAT DISTRIBUTION, REDUCE SPARKS.	2,873,376

cial and important report, TID 5232 in Division 1, Volume 12 on the "Chemical Processing Equipment: Electromagnetic Separation Process" was on one of these microcards. Our chemists had to have the paper report in their hands. Dhafir instructed me to find, hell or heaven, a microcard reader that can print the images, thirty years after the demise of that technology.

Before the onset of the Christmas holidays in December 1987, I sat near the phone in Adil Fiadh's office and dialed the world; I called Japan, South Korea, Germany, France, Sweden, UK and France repeating and endlessly trying to explain in comprehensible terms what a microcard reader should be doing, to the utter bewilderment of my listeners. We tried, at that time, to avoid the US for an additional layer of covert cover. After several hours of phone calls, I finally called Dhafir informing him of the futility of my search and asked for his permission to call the US, our last resort. "Go ahead". I started to call the US. Somewhere in Texas, I got hold of this Egyptian fellow who finally caught a glimpse of what I was talking about.

"Oh, the microcard reader?"

"Yes, yes", I confirmed, holding back my delight that somebody had in fact recognized the word.

"Well, I just came back from an exhibition held in Chicago last week and I think I saw this machine there. Hold on, let me look through my business cards and maybe find the one of that lady that was showing us this really antiquated technique. Are you sure you want this? Nobody even looks at it".

I was sitting on the edge of my seat when he exclaimed, "Here it is. She works for Bell and Howell".

I got the lady just before she was leaving for her Christmas holidays. "How much?" I asked.

"We are selling them for $10,000 a piece, cash in advance".

I inquired further, "How much with three years of recommended spare parts?"

, h, let me see here, that would come to $12,000."

"Ho. J on, please". I called Dhafir even though it was in the
very ea 'v morning hours. "Buy two", he said, and hung up.

"OK, I would like two of these with three years recommended spare parts for each".

"All right, but the price now is $15,000 each".

"You will be getting paid when you come back from your
Christmas vacation, in cash. Just put through the order,
would you, please? And, ινιerry Christmas."

Adil Fiadh took care of the rest of the matter.

By the end of 1987, and within four months of my joining
Dhafir, the scientists and engineers had their hands full of
immediately applicable scientific information on the Calutron
process. They quickly set to work on the Baghdatron.

However, that was just the beginning of the information
avalanche.

Guided by the awareness of the relevance of proper
Quality Assurance and Quality Control procedures that I
had acquired from the several visits a few years earlier to the
IAEA in Vienna while training on the stringent safety procedures required for the site selection process of the nuclear
power plant, I scoured the design offices, drawing rooms and
production floors of the various PC3 departments that were
heavily involved in bringing the EMIS process and the
Baghdatron into being. "The engineers, designers and
machinists are not working according to industrial standards
and procedures", I complained to Dhafir. "This will cost us
heavily in time and will waste tremendous resources".
Furthermore, some of the scientists and engineers were fre-
quently questioned about the numbers of the exact industrial standards that were required for the specification of the

urgently required covert purchases. They did not have at their disposal the context or the numbers of the required standards, especially during exacting time limitations to conclude their purchasing negotiations.

The only repository in Iraq, at that time, of the documents pertaining to the national and international industry standards, as well as Iraqi and Arabic standards, was the Central Agency for Measurements and Quality Control (CAMQC). My first visit to them yielded the temporary lending of the indexes of the leading national industrial standards, American, British, German and the international ones, the International Standards Organizations (ISO) and the International Electric Commission (IEC).

On the spot, I bought five heavy-duty copying machines and assigned a 24-hour shift staff to make copies of these indexes. These copies were distributed to our leading designers, engineers and workshop supervisors. They were asked to pin point the standards that they felt were relevant to their work. With special permission and authorization, arrangements were made with the CAMQC enabling two pickups and a staff of five to visit the Centre every Thursday noon, just before they closed for the weekly holiday on Friday. My staff would gather a good portion of the standards designated by our engineers and production staffs, bring them over to the Nuclear Research Centre at Tuwaitha, and spend the next 36 hours photocopying them. On Saturday morning, with the start of work at the CAMQC, the two pickups would be waiting at the door for my staff to put back the standards where they found them on the shelves of the CAMQC.

Yet we soon found out that many important industrial standard documents were missing from the CAMQC library. By law, CAMQC was the only Iraqi government agency that was authorized to buy the industrial standards from abroad. After a few faltering attempts with CAMQC and getting entangled in their heavy red tape, we sidestepped that law. We leased all of the published standards.

Dhafir had called me to his office. A pile of his weekly quota of Time, Newsweek, Janes Defense Weekly and other periodicals, that he assiduously read every week, were in front of him. He pointed to an advertisement in an up-to-date Aviations Week magazine. An American company, Information Handling Services, which was started by two engineers in 1956, was offering on lease terms a complete set of American industrial standards (issued by more than 200 industrial societies), as well as European and International standards. These were to be provided on about a thousand microfiche rolls; each roll would have about 5000 frames. The library would be updated, by airfreight, every two months. Furthermore, the company offered a complete cross reference to any product available in the US market, providing the complete and updated catalogue of the company that would be selling the product, the specifications of that product, its listed price and the standards that would apply to the product had you intended to manufacture it yourself. Furthermore, another cross reference, utilizing the same indexing scheme, was used to find all the brand names patented in the US, as well as the manufacturing processes for these products where available. In addition, one could also lease the entire American Military standards library, the current as well as the historical ones, on another 1000 rolls of microfiche. The cost of leasing such a total library, along with their sophisticated and automatically controlled microfilm readers and printers, was about $250,000 per year.

Through a middle man in Khobar, Saudi Arabia, we managed to obtain two sets of the whole above microfilm library for the price of one in a matter of a couple of months. Our middleman insisted on bringing with him a representative of the company, as they wanted to be sure that we were the Iraqi Ministry of Industry. Dhafir and I obliged them by

الفعالية	اسم رئيس الفعالية	العنـــــوان
٤أ	د. محمد عبد الزهرة	MOHAMAD ------- SENIOR PHYSICIST
٤ب	د. عبدالله كندوش	QAWAN ----- SENIOR RESEARCHER
٤ج/ر	د. زهير القزاز	NAMIR ----- SENIOR CHEMIST
٤ج/و	د. رياض الجراح	OMAR ----- SENIOR CHEMIST
٤و	د. زغلول نعوم	RUSTAM ----- SENIOR ENGINEER
٤ز	د. غازي الشاهري	PAKRAM ------- SENIOR DESIGN ENGINEER
٢أ	د.ثامر نعمان مولود	AHMAD ······ SENIOR RESEARCHER
٢ب	د.خلوق رؤوف	BASSAM ······ SENIOR RESEARCHER
٢ج	د.باسل الساعاتي	CHASSAB ······ SENIOR CHEMICAL ENGINEERING
٢د	د.فاضل الجلابي	DAGHIR ······ SENIOR CHEMICAL ENGINEERING
٢و/م	منقذ القيسى	FAHAD ······ SENIOR MECH. ENGINEER
٢و/أ	زهير الجلبي	FAHMI ······ SENIOR ELECTRICAL ENGINEER
٢ز	د.حسام الجميلى	GHASAN ······ SENIOR METALLURGIST
٣د	حارث عامر	HASHIM ······ SENIOR DESIGN ENGINEER
٣و	عماد خدوري	IHSAN ······
٣ز	يحي نصيف جاسم	JAMIL ······ SENIOR ENGINEER
٣ط	محمد احمد فؤاد	KAMIL ········ SENIOR ELECTRICAL ENGINEER
٣ح	باسل محمود	LAITH ------- SENIOR ENGINEER
١٠٠٥	د. علا' محمود	TARIQ -----
٢٠٠٥	عبد الجبار الزبيدي	USAMA -----
الاسناد الهندسي	معمر	VAROUSH

جدول رقم ٢ : الاسما' الوهمية للفعاليات في أستخدام تلكسات و بطاقات طلب المعلومات :

A translation of the above list (facing page) of the covert Petrochemical 3 (PC3) Departments names that were used to obtain scientific and technical information follows. The first name would indicate the Department and the last name would be the first name of a staff member of that department requesting the information.

Group 4 (Head: Khalid Ibrahim Said)

Dept	Specialty	Head of Department	Covert Name
4A	Theoretical Physics	Mohammad Abdul Zahra	Mohamad _____
4B	Applied Physics	Abdallah Kendoush	Qawan _____
4C/R	Radiochemistry	Zuhair al-Qazzaz	Namir _____
4C/W	Chemical Engineering	Riyadh al-Jarrah	Omar _____
4W	Electrical Engineering	Zaghloul Kassab	Rustam _____
4Z	Metalurgy	Ghazi al-Shahiri	Pakram _____

Group 2 (Head: Jafar Dhia Jafar)

2A	Theoretical Physics	Thamir Mawlood	Ahmad _____
2B	Applied Physics	Khalouq Raoof	Bassam _____
2C	Chemistry	Basil al-Saati	Chassab _____
2D	Chemical Engineering	Fadhil al-Janabi	Daghir _____
2W/M	Mechanical Engineering	Munqith al-Qaisi	Fahad _____
2W/E	Electrical Engineering	Zuhair al-Chalabi and Basil al-Qaisi	Fahmi _____
2Z	Mettalurgy	Husam al-Jumaily	Ghassan _____

Group 3 (Head: Dhafir Selbi)

3D	Design Engineering	Harith Amir	Hashim _____
3W	Information, Documentation	Imad Khadduri	Ihsan _____
3Z	Mechanical Manufacturing	Yahya Nussaif Jassim	Jamil _____
3T	Electrical Manufacturing	Mohammad Fuoad	Kamil _____
3H	Construction Engineering	Basil Mahmoud	Laith _____

Support Departments

5100	Design support	Ala'a Mahmoud	Tariq _____
5200	Mechanical Engineering support	Abdul Jabbar al-Zubaidi	Usama _____
Support	Electrical Engineering Maintenance	Muamar	Varoosh _____

Location	Fax	Tel	PO Box	Telex	Symbol	Name

((ســـــري للغايــــة))
::::::::::::::::::::

جدول رقم ١ : القنوات المستخدمة في فعالية ٣و

الموقع	FAX	هاتف	صندوق بريد	تلكس	الرمز	أسم القناة
شارع النضال وزارة الصناعة و التصنيع الطابق السابع	8889858	761		2784 ENGSERV	B ٭	مديرية الخدمات الهندسية والفنية وزارة الصناعات الثقيلة MINISTRY OF HEAVY INDUSTRIES ENGINEERING SERVICES DEPARTMENT
المسبح / كرادة خارج عمارة سلمان داود الطابق الثاني	7765448	14206		3512 PERREL	C ٭	مكتب أحمد رشيد أحمد للتجارة والخدمات الفنية (لصالح وزارة الصناعة و التصنيع العسكري) BUREAU AHMED RASHED FOR TECHNICAL AND ENGINEERING SERVICES
شارع النضال		1240		2385 TALEMALI القديم ملغى 2678 TALEM الجديد ستعمل	D ٭	وزارة التعليم العالي و البحث العلمي MINISTRY OF HIGHER EDUCATION AND SCIENTIFIC RESEARCH
شارع النضال وزارة الصناعة و التصنيع الطابق الرابع / غرفة ٧	8874205	8874205 8872006/ 2275 2272	14271	3764 GSD	F ٭	المديرية العامة للتجهيزات الصناعية وزارة الصناعات الثقيلة GENERAL DIRECTORATE FOR INDUSTRIAL SUPPLIES MINISTRY OF HEAVY INDUSTRIES

الموقع	FAX	هاتف	صندوق بريد	تلكس	الرمز	أسم القناة
جادرية	7763250	255 القديم ملغى 29049 الجديد ستعمل 14284 باب المعظم ملغى		3976 SRC القديم ملغى 4218 CSCRE الجديد ستعمل 4244 SRCLB مطبوعات حساسة	ب ع ٭	مجلس البحث العلمي SCIENTIFIC RESEARCH COUNCIL
شارع السعدون مقابل محطة البنزين	7183250/ 146 هدى	198		2230 SCOP	ش ن	شركة المشاريع النفطية STATE ORGANIZATION FOR OIL PROJECTS الاسم : AL-SAMMARI
الوزيرية مقابل الجامعة المستنصرية	4253445	11118		4058 ELIND	ص هـ ٭	المنشأة العامة للصناعات الكهربائية STATE ELECTRICAL INDUSTRIES ESTABLISHMENT الاسم : علي محمد علي
المصافي	7750280	523		2649 DISDAR	م ن	شركة المعدات النفطية OIL EQUIPMENT COMPANY الاسم : عاطرة
شورجة مقابل جامع الخلد	8885141			4001 BED	ك ب ٭	المنشأة العامة لتوزيع كهرباء بغداد STATE DIRECTORATE GENERAL OF BAGHDAD ELECTRICITY DISTRIBUTION MINISTRY OF HEAVY INDUSTRIES الاسم : يوسف عمانوئيل يوسف
الزعفرانية		29042			ص ن	الصناعات الالكترونية ELECTRONIC INDUSTRIES CO. MR. F. TAPOU/IMPORT MANAGER
		5815		2182 MUDT	م ف	مديرية المعدات الفنية MINISTRY OF INDUSTRY AND MINERALS DIRECTORATE OF TECHNICAL EQUIPMENT

٭ : جهاز التلكس متوفر في المنظمة

The "channels" through which Petrochemical 3 (PC3) requested its scientific and engineering literature and used for its covert purchasing of materials and equipment, 1988.

٭ 110 ٭

meeting them in the best boardroom of the Ministry of Industry in the center of Baghdad and took them on a tour to the intended location for the microfilm library, freshly decorated and staffed.

Khitam Kathim, a recent high school graduate, with a keen sense of learning, gracious calm, shy mannerism and endless patience was assigned to look after the library. She so mastered the technique that she could provide, within a time span of half an hour, a scientist or engineer searching for a product, all the companies that offered the product, its stated price, the standards that applied to it, as well as the actual standards themselves, in hard paper copy. The same applied for any US military item. A few months later, the American company provided us with a similar service for European and Japanese products. Our covert purchasing simply took off.

Again, I did the rounds on my fellow scientists and engineers. This time I had with me about fifty scientific and engineering monthly publications. I would be willing to provide, for the chosen few, a regular and steady supply of these periodicals that matched their field of specialization, on one condition. Each of these fifty periodicals had a card filled with numbered boxes, about 50 to each card. All of the articles, advertisements, conference calls and book reviews that were mentioned in the publication would have a small number next to each item. The reader would only need to check the desired numbered box on the card and mail it. The publisher would provide, free of charge, the required information, be it a catalogue of a product, a conference call details, or a complete reprint of an article mentioned as a reference. Our condition stipulated that the scientist or engineer had to hand in a checked card once a month; otherwise, I would take the periodical from him and give it to an eagerly awaiting colleague. Initially, these periodicals were not allowed to be taken out of the Atomic Energy Library, but had to be read and remain in the Library, which limited their use. To give them out for keeps to the scientist and engineers, and circumvent the tight auditing and inventory control procedures laid down by Dhafir, was no small feat. Jafar had to issue the order that I could subscribe to these 50 periodicals, but that no tracking record should be kept for them once they arrived and that they may be retained by the selected fifty.

The cards started to flow in. I devised a system of pseudonyms matching the various departments in PC3. Khawla, the almond-eyed beauty, Nesreen, a most efficient strong character woman, and the ever serene and loyal Loai, Salam's friend, would keep track of who submitted their cards. Armed with pads of stamps, they would record the traffic of posted cards and send them on their way. Twice every week, the two pickup trucks would make the rounds to more than a dozen post offices, each registered under a different legitimate government agency name, to collect the pouring information. We even left special collecting containers for the mailboxes that were overflowing their capacity. Twice a week, three to four overflowing packed mailbags would be hauled to my office. I would spend the whole night opening each envelope, decipher the correct person it was addressed to, and glance through or read the technical and scientific literature destined for him/her. In this manner, each of the fifty recipients managed to build up, within several months, an up-to-date technical and scientific library in their own office and under their total command for reference. The cost was a postage stamp a month for each.

This particular information venue catapulted me into a focal role. Whenever we held our regular brain storming scientific and technical sessions, and reached a point requiring an information input, such as, for example, the need for flexible stainless steel bellows or oxygen free copper tubes, Jafar would pick me out of the audience and I would point to the engineer or the scientist who would know about the specification or have access to further information about that particular subject. I knew the contents of each of my fifty library keepers. I did not miss any brain storming sessions.

Since the mid-seventies, I was already in charge of accessing Dialog, the world's first online information retrieval system, created in 1972 by Roger Summit, from the Nuclear Research Centre. We had a special small isolated room on the outskirts of the Nuclear Research Centre equipped with a dumb terminal, modem, long distance telephone line, a line printer and one chair. I had the only key to that room. I would be provided with the key words to be searched, such as plutonium. Inventiveness was required to devise other key words that were safe to ask but might still yield information on the sensitive key words required. It was expensive to have an open line from Baghdad to California using a dumb terminal and a slow modem. By the mid eighties, there were about 600 available databases; a few of them, such as the aerospace databases, were restricted to searches from the US, Canada and certain NATO countries only. Dhafir arranged for a station to be set up in Madrid, Spain for me to go and access Dialog's restricted databases from there.

In order to secure the supply of special books, reports and hard to get articles, several accounts were opened with various information suppliers. These included the British Lending Library, the Institute of Electrical and Electronics Engineers (IEEE), the University of Michigan's UMI (University Microfilms International), Blackwell, a book supplier in London, and Micro Info in the UK. Adil Fiadh saw to it that the ten thousand dollars deposited in our accounts at each information supplier would be promptly replenished every year. A stage was reached when we were able to get via airmail any book, magazine, article, patent, report or periodical within one week of its request by our scientists or engineers, by a simple phone call or a fax.

A final information chore was Jafar's assignment for me to liaison with top-level scientists and engineers who were working in universities and other government agencies but whose skills and knowledge were eagerly needed in PC3. This undertaking required a lot of tact and good will. My colleagues would point out such an expert to me. I would then pay the intended party a personal visit at first and explain the kind of work that is required to be performed, the expected scope of results and the estimated time span as well as the generous remuneration for the work. Their departments were not necessarily informed of this arrangement. More than a dozen such assignments were implemented, ranging from the design of the especially designed electric substation that was required to run the bank of Baghdatrons at the Tarmia site intended to absorb and dampen the huge electric spikes of the fast switching magnets, to the design of especially formed high explosive devices.

By the summer of 1990, the scientists and engineers at PC3 were flooded with pertinent information for their work or available on demand in short notice. I had spent about half a million dollars in three years. A favorable report was submitted by Jafar in the summer of 1990 to Hussain Kamel on these activities and pointed to the wealth of information that my department had managed to accumulate and master. Jafar proposed to disseminate these potent services. Hussain Kamel concurred and ordered that my department should surface to the 'public" domain and to be henceforth of service to all Iraqi ministries, research centers and universities, free of charge. My department was the first department from PC3 to become public. All of the other departments, according to Hussain Kamel's orders, followed suit after the war in 1993 resulting in the total demise of the nuclear weapons program. PC3 was by then dispersed into more than a dozen civilian enterprises that reflected the technical specialties of the concerned departments. My department, 3W, except for Khitam and her copy of the microfilm library of standards and company information that remained in the cellar of the Atomic Energy library, moved to the Ministry of Industry in the autumn of 1990. Khitam joined us after the war and after the bombing of Tuwaitha. The Centre for Specialized Information opened its door as a free information provider in October of 1990.

Documentation

The above information gathering activities was only one part of my responsibilities in PC3. The other part was documenting the scientific and technical reports that were pouring out of the work of the various departments at PC3, their archiving and their hiding before the 1991 war.

Having laid down the original documentation procedures in 1981 and 1982, the responsibility was handed to a couple of others after my departure in 1983. Upon returning to this activity in September 1987, it was not a simple task to regroup the documentation team under the able, dedicated and meticulous Hamid Kathim. We did our best to recuperate the documentation process from some of the previous lapses that had resulted in erroneous numbering, logging and distribution. In any event, the TIG and PIG enrichment processes were being abandoned and a new full drive toward EMIS was being launched. I thoroughly revamped the documentation procedure taking into account the new organizational structure of PC3, and ensured the presence of the complete set of the program's reports in three physically different locations, a chore that had lapsed during the previous few years. The originals were kept in Building 61 at the Nuclear Research centre, which was the Electronics Department under Basil al-Qaisi. The second set was at the Trade Union building in front of al-Rasheed Hotel, the location that was targeted by David Kay in September 1991. The third location was al-Hayat building, an Intelligence adjunct near the presidential palace. Hamid and his staff of ten labored dutifully, at the basement of building 61, in maintaining the records, making microfilm copies of the engineering drawings and producing the required number of copies of the reports to be distributed to scientists and engineers. They also carried out a yearly physical audit of every issued copy in the possession of everyone, including Jafar and Humam, the Heads of PC3.

No PC3 report, to our knowledge and on hindsight, did leak into other Intelligence agencies' hands until after the 1991 war.

The main responsibility of this task, however, was not only to ensure the scientific quality of the submitted research reports, but to also document the reports submitted by senior scientists and engineers upon their return from assignments abroad either to attend scientific conferences or from covert purchasing sprees. After careful review, many of the reports were returned back to their originators for further refinement, additional work or more useful detail. Only when that was satisfactorily accomplished would I submit the report to Jafar for his final approval. A recommended distribution list would also be submitted for Jafar to edit and refine.

In 1988, a storm brewed with Khalid Said that had a disastrous sequel in 1991. Khalid had refused to have Group 4's scientific reports be incorporated in my newly revised documentation procedure for PC3. He insisted that Groups 4's reports documentation be completely separate and under his own exclusive domain, with his own numbering, distribution and archiving. They were to be stored at the al-Athir site, the headquarters of Group 4, about 30 kilometers southwest of Baghdad that was being completed. I vehemently opposed that proposal. Bitter exchanges, to put it mildly, ensued. I stood my ground. There had to be one central repository for the nuclear weapons program's reports, sans Group 1, which was physically and administratively separate and under Hussain Kamel's jurisdiction. After several weeks of heated debate, a compromise was reached. Khalid would submit to me an abstract of the reports, to be numbered and indexed in our ledger according to our procedures, but that his documentation group, at al-Athir, would be in charge of their copying, distribution and safekeeping.

While on the subject of the al-Athir site, it is appropriate to quell yet another false claim of Khidir Hamza. He did boast a great deal in his book of his own role in building the al-Athir weapon manufacturing centre during the late 1980s, while in fact he was a mere has-been doing precious little at the Tuwaitha Nuclear Research Centre. He retired in 1989 when al-Athir's activities were peaking. He did not even have an office space in al-Athir, and I doubt very much that he ever set foot on that site.

In early 1990, my attention was drawn to the capabilities of a promising archival apparatus that was destined for release by Canon, the Japanese company, in the summer of 1990. CanoFile 150 was a scanning machine that could capture the image of both sides of a scanned document and store both images simultaneously on a magneto optical disc. It could scan and store the images of up to sixty pages per minute. Each magneto optical disc could store up to 10,000 images. I asked Canon's representative in Baghdad to allow us to take a look at this device. It was hurriedly shipped to Iraq in the spring of 1990. Salam, his friend Loai and I were trained on its use at the Ministry of Industry. Indexing of the scanned documents was of prime importance in order to be able to retrieve the individual reports. I did point out to our Japanese salesman, with a smile, a fine printed label on the side of the device advising the buyer that "this is a strategic and sensitive device requiring a special US export license", noting that it was made in and sold by Japan. We bought the first device that was shipped out of Japan in June 1990 and ordered a second one, along with three years of consumable spare parts and five empty disks. Canon's representative offered the sixth disk, on which we did our training, as a gift. I put that one aside, as I am known for harboring backup contingencies. The second device arrived with the last plane from Japan on the night of August 2, 1990, the night that the Iraqi troops entered Kuwait. Air travel was immediately suspended the following day and for the next thirteen years.

With both devices ready, and the war clouds gathering, the whole documentation staff set to work scanning and saving the 1600 reports that had been accumulated as a result of ten years of work and development. Khalid Said was also coaxed to send his documentation staff with their reports. They were given a separate empty disc to store their reports. Once done, I refused to give that disc back to Khalid, as things were getting tense and a new demand was forthcoming: We had to hide the archived documents.

Salam and I went to the bazaar near al-Mustansiryah Street and bought three large aluminum trunks. Having often visited the Tarmiya site housing the EMIS process under Dhafir's dynamic management, and where Dhafir and I would enjoy an occasional quail hunting trip in the surrounding area, a newly German built secondary technical school, at about seven kilometers away from the site, caught my attention. Salam and I toured the technical school and we found a windowless room that could only be accessed, for some odd reason, by going through two other rooms. That would be the place to hide the reports of the Iraqi nuclear weapons program.

I pleaded with Dhafir to not let it be known, especially to the Intelligence and Security people, of when and to where Salam and I would be moving the filled aluminum trunks. I did not especially trust the juniors in those organizations. Reluctantly, he agreed. Salam and I managed alone. The officials at the technical school welcomed us as members of the newly created Centre for Specialized Information from the Ministry of Industry. A few days later, I announced to Dhafir that the trunks were safely deposited and handed him one set of keys to the newly changed heavy locks of the three rooms. Salam and I kept the only other two sets of keys. I still hung on to the magneto optical disks that were laden with the images of 1600 PC3 reports.

A week later, I was thoroughly enraged. Khalid Said and his documentation group had finally relented to hide their reports along with ours. Their messenger had simply dumped a cardboard carton full of reports on my desk and left, with no signature, no acknowledgment and no warning. I was visibly furious over this irresponsible and belated action. Utterly despondent, I dispatched Salam to take the carton to the secondary technical school. As the aluminum trunks were fully packed and accurately indexed, Salam, undoubtedly also very upset, did not bother to rearrange things yet again and simply left the cardboard carton on top of the trunks. Fate had it that the infamous David Kay would lay his hands on that same cardboard box a year later. This directly resulted in the blowing up of the al-Athir site by the UNSCOM inspectors, under the orders of Zifferero who was one of the heads of the IAEA inspection teams at the time.

As mentioned earlier, our vast holdings of scientific and engineering information was deemed by Hussain Kamel to be of service to all Iraqi government agencies, universities and research centres. Hence, we moved to the mezzanine floor of the Ministry of Industry in the autumn of 1990 and set up the Centre for Specialized Information. My deputy at department 3W, Mashkoor Haidar, was to take over the documentation responsibility from me for PC3. Mashkoor was not senior enough. Hence, Mashkoor and the staff of the documentation group were assigned to Adil Fiadh. I handed Mashkoor the keys to the three rooms, but hung on to the magneto optical disks. Jafar was upset and demanded that I hand those to Abdul Halim al-Hajjaj, Khalid's associate. I strongly objected and argued fervently that they would be safe with me at the new location, the Ministry of Industry, where I had taken one of the CanoFile archiving machines. Jafar's tone of voice rose, which was a very rare occasion. With a broken heart, and spirit, I handed over the three full magneto optical disks to Halim. Jafar was still looking for them seven years later.

War was approaching.

Selected senior scientists, as well as the management team, were assigned alternative domicile in case of hostilities. The al-Fajir (the Dawn) site, which was built near the ancient city of Sharqat, about 200 kilometers north of Baghdad, was a replica of the al-Safa (The Tranquil) site at Tarmia that housed the EMIS process and the Baghdatrons. A housing complex for hosting about fifty families was built about 7 kilometers north of the actual replicated site at Sharqat. The able and hospitable Mouwafaq Matloub was in charge of al-Fajir. Situated among rolling hills and overlooking the distant Tigris River, the construction had reached very advanced stages and was nearing completion. I understood that not a single foreigner was employed in the vast construction effort, due to its secrecy.

On a Wednesday morning, the day before the bombs started raining down, I drove to Sharqat to lay claim on one of the houses there and to determine what else was needed to be brought with us. The fog became so thick that day that I could not even see the white stripes along the side of the highway. Pulling over on the gravel, and as far away as possible from the road, I waited for an hour for the fog to clear. Back in Baghdad, I spent the afternoon with Salam Toma securing the contents of the Centre for Specialized Information. All the computers were covered with plastic sheets and stored in the

basement. Most of the portable trays of microfiche and catalogues were taken to our homes. We dispersed the racks of
microfilms in different locations so as not to suffer from a single hit. At home, personal suitcases were prepared, official personal papers gathered and dozens of batteries were purchased.

At 2:30 in the morning, the horrid wailing screams of the
sirens tore through the night, cold and demon-like in comparison to the heart felt wailing of old women at the death of
a family member. I pulled our enthralled children, who were
watching the death bringing fireworks, away from the windows, lest a wave shatters the glass in their faces. With electricity and telephone systems down, and in the faint light of
dawn, we packed the two cars; Niran's and my government
assigned one, and drove off to Sharqat with my mother and
Lisa, our dog.

A foreboding thought kept haunting me all the way to
Sharqat. What if the "smart" Americans bombs missed their
target, breached the containment of the Russian reactor, and
released devastating radioactivity? A mini Chernobyl would
have afflicted Baghdad. In taking such a blind risk and bomb
the Tuwaitha Nuclear Research Centre relentlessly, the
Americans have maintained their legacy of Hiroshima and
Nagasaki in Japan and Agent Orange in Vietnam.

It transpired that the bombs did in fact fall on Tuwaitha
while the reactor was still operational. The operators first
fled the building when the bombs first fell close to them but
then bravely returned, shut down the reactor and put a steel
cover over the open pool as the bombs exploded tens of
meters from the building. Fortunately, that steel cover was
not breached neither was the concrete containment of the
reactor holding the water that cooled the reactor.

The Iraqi nuclear weapons program stopped dead in its
tracks that morning, and never rejuvenated.

How close was Iraq to obtaining a nuclear bomb after ten
years of its program?

The team at Tarmiah under the guidance of Jafar and Dhafir had managed to collect, at most, about 5 grams of weapon grade uranium 235 with the Baghdatrons. The core of the bomb, along with its casting, would have required 18-20 kilograms.

The actual design of the bomb among the Group 4 departments at al-Athir was still under consideration and by no means frozen on any final design. There still remained some scientific considerations regarding the total weight and a few tests of the extremely accurate electronic explosive triggers that would form the shock lenses that would implode the uranium core into the right density to sustain a nuclear chain reaction. A preliminary investigation was just beginning to find a site in the desert to test the bomb, once it would be readied. This test would also have doubled the amount of the required weapon grade uranium. Finally, the delivery and guidance systems were still being considered and not fully developed.

In total, we were, in my estimate, about 10-20 percent of where we should have been had Iraq had a nuclear weapon. It would have required further several years.

CHAPTER SIX:
Disintegration and Escape

Through the 1991 war

Though provisioning was well catered for in the housing complex at the al-Fajir site in Sharqat under the warm and watchful eyes of Mouwafaq Matloub, all semblance of a normal functioning society outside was disappearing quickly. The power stations' electrical grids were covered with air dropped special nets embedded with graphite pea sized pellets that caused extensive electric shorts bringing the whole electrical distribution over Iraq to a halt and hurling the whole country into darkness save for those with mobile power generators. The telephone system went down during the first day of the war.

I returned to Baghdad several times during the war to inspect the status of the Centre for Specialized Information. Reaching Baghdad during the war was a harrowing experience in itself, with bombs exploding along the road, bullets strafing the highway and just enough petrol to get us into Baghdad. The fireplace would be the only source of light and heat as I inspected our home. As night fell, I would ride my bicycle to visit close relatives and friends. Eerily cycling over abandoned bridges and empty roads, I witnessed the streaking anti aircraft fire crisscrossing lines of light in the sky followed by the bright flashes of bombs and missiles destroying their intentional and unintentional ground targets. While staying with my friend Munthir Shammas in Jadiria, we shook as the bombs felled the elegant hanging bridge over the Tigris nearby. Sleeping next to my drunken cousin Salah al-Saigh, I would hear him mutter curses at Saddam even in his sleep.

News filtered to us of discouraging premonitions of some of the war atrocities. A housing complex similar to ours and belonging to al-Badir electrical establishment, which was about 100 kilometers south of Sharqat near the ancient city of Samara, was bombed by the Americans a week into the war killing 50 women and children. Similar news reached us of more civilian casualties in an attack on the Ishtar housing complex near the Tuwaitha Research Center which had previously housed the French contingent that were building the French reactors, and now occupied by Iraqi scientists and their families. Many nights, our house would shake from shock waves that we could at best attribute to ammunition depots exploding kilometers away. My family made often use of the primitive underground shelter that we dug in front of our house. Despite the hard times, there were other comradely events. A fond memory is that of Dhafir and his two brothers cooking a couple of fish, Masgoof style, in front of their house as war planes flew above and bombs fell around us in the distance. Niran and I spent many socially pleasant evenings playing cards with the families of Sabah Abdul Noor and Mahir Sarsam, two senior scientists in Group 4. Late at night, we would carefully trudge back to our home with a lantern shooing off vicious stray dogs.

Sharqat itself, an ancient village that reflected its deep historical roots in the faces of its inhabitants, was spared. It was surrounded by archeological sites and ruins that we often visited. A bus would take us daily to shop for food and other necessities. We fared relatively well in comparison the horrendous onslaught taking place all over Iraq.

Early in March 1991 and towards the end of the war campaign, I went once again to Baghdad to inspect the Centre for Specialized Information. The al-Safa site in Tarmia, where Dhafir was in charge, had been bombed two days earlier. On the way back, and as we gathered in the damaged Tarmia site to catch the bus back to Sharqat,

My family in front of the "underground shelter" at Sharqat during the 1991 war.

Dhafir took me aside and privately informed me that the al-Fajir site in Sharqat was bombed the night before and that he has no information on the extent of the damage or whether the housing complex itself was hit. Dhafir's families, as well as the families of the most senior managers, scientists and engineers were there. I tensely sat quietly among the dozen passengers in that bus returning to Sharqat not daring to reveal this sad news lest I would shatter their calm for the duration of the trip. It was one of the longest three hours in my life. As the bus veered around the last corner heading to the housing complex, I ventured to catch the attention of my fellow passengers and broke the news to them, in order to prepare them for whatever might confront us. Luckily, the housing complex was spared but the al-Fajir site itself, seven kilometers away, was destroyed. By next day, all of the families had left the housing site except for my family and Muwaffaq's. Niran and my mother were of the opinion that since the children took the attack in staid stride; it was of no use going anywhere else.

Hostilities ceased a day or two later and Dhafir came up to al-Fajir for a short visit. I accompanied him back to Baghdad for an assignment. The highway passed 100 meters away from the expansive Baiji oil refinery plant that was situated about 50 kilometers to the south of Sharqat. We witnessed huge 100-200 meters high walls of fire raging from spilt oil that was engulfing the bombed refinery. We both revisited the same refinery a couple of months later to supervise its rehabilitation by our nuclear scientists and engineers. Seeing the extent of the damage grotesquely outlined by the twisted, bullet riddled and burnt structures, I expressed my doubt whether we would ever be able to get it functioning again soon. Dhafir was more in touch with the capabilities of the teams of engineers and highly skilled technicians under his direction. He simply stated, "Inshallah (God willing) within three months". He was right.

A few days after the end of the war, and as we rode the bus to Baghdad with our bundled effects, I turned to Mahir Sarsam. He was a prominent physicist and a close friend who was assigned the task of locating a test site for a nuclear weapon test. I direly predicted, "Allah Yustir (God protect us)", anticipating an instinctive violent reaction of the Iraqi people to our abject defeat. Little had I known of the Basra uprising that was taking place just then and the smashing of Saddam's picture there that had signaled a more widespread revolt among the Shiites in the South of Iraq. Only a year or two later did I learn of the extent of the brutal repression inflicted by the Ba'athist stalwarts on the revolting people, the heroic popular extent of the uprising, the extensive damage to the holy Moslem shrines in Karbala and al-Najjaf and the horrendous mass grave yards. We did hear at the time of the anger at the Americans who had allowed the helicopters of the Republican Guards to fly freely and participate in that repression. The Kurds in the North, like the Arabs in the South, had naively believed Bush senior's call for an uprising, only to be let down, left unaided and be slaughtered. Coming down from Sharqat, we saw some of the Republican Army's modern tanks heading north, unhindered, to quell the Kurdish uprising.

Two weeks after the war and as we stood chatting in front of our homes, our beautiful neighbor, Zainab al-Baia, uttered an ominous and remarkably foresighted prediction. She said that since the uprisings had failed, then Saddam would get much stronger and remain for many years to come. She was only 18 years old. Saddam's reign lasted for exactly another 12 years.

Anticipating that the worst was yet to come, I managed within the first week of our return to Baghdad to obtain fresh new passports for my wife and children. However, being part of the nuclear team, my name was on the list of those forbidden to obtain a passport except for official business and with the knowledge and approval of the Intelligence Agency. That was the start of our long and painful ordeal of secretly escaping from Iraq, seriously considered in 1995 and finally managed in 1997-1998.

As soon as we returned to work in March 1991, I submitted a request for my retirement to Jafar Dhia Jafar, the scientific head of the nuclear weapons program. He was perhaps more bemused than surprised, if not downright incredulous. Being on close personal terms with him, I explained to him the decision that I had taken during the mid-sixties that I would only marry an Iraqi, have my kids raised in the warmth of the Iraqi Arabic culture and traditions, but that I would strive to provide them with the best education that I and my wife would be able to muster, as this was a tradition in my family. I explained to him that I was well aware of the fact that a favorable consideration of such a request would take many years of bureaucratic processing. However, within six or seven years, my eldest children would be reaching university age. Hence, I was aiming for that time period to make sure that my children did get into reputable universities when they finished high school. Politely and diplomatically, he did not refuse the petition but, with a wry smile, scribbled for his secretary to put the matter under consideration for the time being. He had a very serious task for me to accomplish: convening the first Electricity Rehabilitation Symposium in Baghdad, within one week, to assess the damage and rehabilitate the electric power sector. It was just one week after the end of the war.

Rehabilitation

With the eager assistance of my confidant Salam Toma, our librarian Nahida, the information retrieval specialist Khitam and a couple of other colleagues, we resumed work at the Centre for Specialized Information within a week after the end of the war. The assignment given to me by Jafar was daunting. I had to convene a symposium bringing together the top engineers and heads of departments of all the electrical power stations in Iraq, with no telephones to communicate the event and precious little electricity. Zaghloul Kassab's energetic communications group had already established a mini sun powered telephone exchange to connect only our own various departments in Baghdad. I dispatched Salam with an invitation in his hand to the various bombed electric power plants throughout Iraq.

The symposium took place as planned over a period of three days in a theatre next to the Information Ministry and in front of the Melia Mansoor Hotel where all of the guests were freely hosted. The efforts of my staff shone in the level of organization and efficiency, with our limited resources, that led to the success of the gathering. The 150 page proceedings of more than twenty papers that were presented at the symposium were distributed to the participants within a week after its convening. The initial disbelief of the electricity sector's senior staff, who strongly doubted any quick restoration to the electric power facilities, was slowly dispersed as they witnessed the heightened discipline and focused organization of our scientists and engineers, and their total commitment to rebuilding their bombed structures. We were primed for work when the nuclear weapons program had stopped in its track as war broke out. Our efforts were aided by the plentiful stocks of spare parts in the electric power plants' warehouses. These stores were usually situated relatively far away from the main plant structures, and hence escaped the effects of bombing. The abundance of these spare parts was an uncalculated benefit of the standard legacy in pre-war purchasing procedures that required the ordering of three years spare parts for any purchased technical equipment and apparatus.

As the fervent recovery work at the power plants took place and gathered astonishing speed and momentum, we managed to organize two more symposiums during the next three months to follow-up, evaluate and redistribute the rehabilitation efforts. Consequently, and as the electric power flowed once again, the scope of rehabilitation extended to the oil sector and our engineers and skilled technicians tackled the bombed refineries of Baiji to the south of Sharqat and al-Dawra near Baghdad, as well as to the bombed telephone exchange buildings. A third of Iraq's electric power supply was re-established within four to five months. The oil refineries were up and running again in a similar span of time.

The rehabilitation effort was, by any modest standard, a proud proof of the true grit and resilience of the Iraqi spirit that can be wielded by a tight-knit organization structure.

In the meantime, the UN inspectors were to arrive. A memo was written in April/May 1991 by Jafar Dhia Jafar and Naman al-Niami, a top level chemist in the nuclear weapons program, to Hussain Kamel outlining all of the nuclear sites and activities. The list was submitted even before the adoption of Resolution 687 (1991) by the United Nations Security Council. Hussain Kamel ordered the disclosure of selected activities and sites and the concealment of the others from the list, including al-Athir weapon design center and its activities.

Dhafir Selbi had submitted a request for his retirement shortly after the end of the war. Dhafir had heard of my feeble attempt to secure a retirement from Jafar. He approached me with a tantalizing offer. Dhafir had a journalistic streak in him and was (probably still is) very fond of putting his thoughts into writing, sometimes in too abstract, choice words and in a very tight-knit form for my taste. He was encouraged by the news at the time that a law would be enacted allowing the publication of independent newspapers. He proposed to me that he would take up the matter of my retirement to the authority higher than Jafar, namely to Hussain Kamel himself, and attempt to process both his request and mine at the same time. In the event that he did succeed in clinching the retirement for me, I would agree to collaborate with him in setting up his independent newspaper and look after the computing and technical side of the venture. I readily agreed, still very impressed by his honed managerial and intellectual skills. Six months later, Hussain Kamel agreed to Dhafir's request in September 1991 but unfortunately shelved mine, as he was well aware of the potential benefits of the resources of the Centre for Specialized Information to the post-war rehabilitation activities.

Witnessing the slow return of telephone communications in Iraq, I had embarked, in the summer of 1991, and upon my own initiative, on networking all of the research centers and universities throughout Iraq. Employing external Hayes modems, the ever pleasant and capable Ayad Muhaimid and I would, over a period of two years, visit and connect about sixty research centers and universities with a telephone dial-up service to allow them access to the many databases on CD-ROMs that were located at the Centre for Specialized Information in the Ministry of Industry in Baghdad. We were still able to update and purchase new databases through Adil Fiadh's contacts in Jordan. From their locations throughout Iraq, the scientists and researchers would be able to do a full text-based search in the databases on our CD-ROMs and print their findings directly on their local printer. We had accumulated about twenty scientific and engineering databases, including all five million US patents, the entire textual PhD thesis holdings of all American universities and many international ones since the thirties, as well as PhD theses abstracts extending all the way back to 1864. This was on top of the microfilm treasure of industrial and US military standards and industrial catalogues.

Within a few months after the war, we would normally open our offices at eight in the morning to a waiting line of twenty to thirty government engineers, students and university researchers eager to get information, for free, for the rehabilitation of their sectors or for writing their theses.

In one instance, an army staff researcher came from the al-Kindy military research centre at Mosul, in the North. He held in his hand a small piece of hard rubber that was salvaged from a downed American plane. The researchers at al-Kindi wanted to know the process for manufacturing a similar material. After examining the piece of rubber, I noticed a small imprint of its brand name on one side. Referencing our brand name catalogue for that imprint and cross-indexing it with the company that held the patent for that brand name, a fax of 50 pages was sent to al-Kindi outlining the complete manufacturing process. The whole affair lasted within half an hour from the visit of their messenger.

Our librarian, Nahida, trained more than 400 users on the use of the telephone-based network that Ayad and I had installed throughout Iraq. The staff at the centre numbered no more than ten. I gave them sufficient authority to run their assignments, and had instructed them to refer back to me only when there was a serious problem. We even wrote our own computer program to distribute our monthly salaries, as the department responsible for that in the now slowly disintegrating PC3 was incapable of running their own program on the relocated and dismembered mainframe computer. This unorthodox mode of independence and open style of management did not go well with the strict security measures preferred by the apparatus of the Military Industrialization Corporation (MIC) to which we were loosely appended, as events unfolded two years later when we finally did fall completely under their authority.

Hussain Kamel's internment

The UN inspectors finally started to arrive in the summer of 1991 and commenced to interview some of our nuclear scientists and engineers who had to be diverted from their rehabilitation efforts. They raised a troubling complaint to Jafar. Many of them requested access to their reports in their discussions with the UN inspectors. Jafar, by then, was appointed Head of the MIC, still under Hussain Kamel's authority, having led such a successful rehabilitation campaign of the electricity sector. The scientists and engineers convinced Jafar of of their case. It was decided to hand over the contents of one documentation centre to the UN inspectors. These should have encompassed the reports of the declared activities only. In late summer of 1991, Jafar gave a fatal order to Adil Fiadh to bring back the concealed documents and reports.

Adil Fiadh had assumed responsibility of the reports that were hidden in the secondary technical school after my move to the Ministry of Industry in the autumn of 1990. After the end of the war, he had instructed my ex-assistant, Mashkoor

Haidar, to remove the complete collection of documents from the sealed room and put them in a freight train wagon, with welded doors, that kept traveling between Basra in the south and Mosul in the north. Upon getting Jafar's order to return the documents, the train was halted and the welded doors pried open. The aluminum trunks, boxes of microfiche of design drawings and the cardboard box containing the reports of the undeclared activities of Group 4 (that were belatedly dumped on my desk) were all returned to the documentation centre at the Labor Union building, next to the MIC building.

Both Salam and I were by then outside the information loop of what was happening to the documents and, therefore, could not offer our feedback in light of the policy of "declared" and "undeclared" activities and its repercussions on the mixed content of the documentation trove. No other timely warning to this effect was raised in the confusion and myriad of activities.

Within a few days later, the UN inspector David Kay and his colleagues unexpectedly raided the Labor Union building and laid their hands on the documents, including the cardboard box, leading to heated verbal exchanges and face-to-face confrontation between David Kay and Jafar, which was videotaped and broadcast. A week later, the inspectors raided the al-Khairat Building in Sa'adoon (situated in front of the now famous Meridian Hotel and near the Firdaws square where Saddam's statue was toppled by the Americans) which was the temporary location of the PC3 staff, scientists and engineers. There, they found even more documents and detailed computer information on the personnel and activities of PC3.

Hussain Kamel suspected a security leak. He ordered the arrest of about twelve people connected with documentation, Adil Fiadh, Mashkoor Haidar, and I included. We were each interrogated by a committee headed by the Deputy Head of MIC, Amer al-Ubaidi (who later became the Oil

Minister in 1996 and was captured by the Americans in May 2003). Though Amir and I were both students at the University of Birmingham, he angrily and insultingly reprimanded me for bringing in my pipe to the interrogation session. We were interned, incommunicado, at the Fao Establishment building on Palestine Street for eighteen days. Some of the interned suffered psychologically, broke down and cried heavily, realizing that our lives were at the whim, ever so fragile, of Hussain Kamel's mood.

From the fifth floor window where we were kept under constant guard, I would be able to see Niran's car in the morning as she went to work. She had no idea of where I was. Only the comforting warm gestures of Ihsan Fahad, a senior PC3 scientist entrusted with a managerial position at the MIC, kept our hope at a sustainable level. It finally became apparent to Hussain Kamel that none of the interned was at fault and that there was no security breach. He released us after demoting all of us administratively, including Khalid Said. Jafar was removed from the helm of MIC with Amer al-Ubaidi ascending, yet again, to that post. Jafar was assigned a recluse post of "a scientific advisor" to the palace court of Saddam and assigned the full responsibility for the continued rehabilitation of the electricity sector. Jafar also dabbled in fanciful irrigation projects to divert water from the Tigris near Mosul to irrigate the fertile desert near the ancient town of Hathar about 100 kilometers southwest of Mosul. I was abreast of all research activities that were taking place in PC3, and some of the military research activities of MIC, since the researchers had to pass me to get to the troves of information at the Centre of Specialized Information. That held true until 1994 when I left for the Foreign Ministry.

Nevertheless, and despite our release from Hussain Kamel's internment, the Intelligence Agency maintained their surveillance vigil and kept an ever-watchful eye over my daily activities, contacts and friends. It attempted to recruit several very close associates and friends of mine for this task. Some of their contacts, like my faithful secretary Salma, were brave enough to inform me of their assignment to her soon after she was approached. She filled me in on the minute details of her contacts with them at the risk of her own peril, and handed them bland information about my activities. Others, like my loyal Palestinian friend Husam Obaid, who partnered in our small computer shop for eight years during the nineties, would only reveal, after our escape in 1998, his steadfast refusal to the repeated Intelligence attempts to recruit him to spy on me. Husam suffered at the sight of my depression and stress that lasted for years, as he watched me painfully pace the sidewalk in front of our shop, in deep and anguished thought. He revealed to me that the Intelligence Agency's concerted attempts to force him to spy on me was the reason that had forced him to opt out and leave Iraq in 1995, claiming that he was seeking work abroad, but it was in fact so as not to succumb to their threats. He did return to Iraq several months later and managed to solidly resist their attempts to recruit him until my escape, which thoroughly infuriated them. I had simply kissed Husam the afternoon of our departure. He immediately sensed what was about to happen and bitterly cried as we hugged, not knowing how we would escape and fearing for our fate in case we failed. The day after our escape, he was dutifully visited by the secretary of a senior Intelligence officer and was threateningly treated. When his interrogation passed the psychological threshold and verged on physical harm, Husam simply packed up one night and left Iraq.

I am sure that there were other informers, close to me, that I am not yet aware of, who had monitored my movements and contacts and thickened my file at the Intelligence offices.

Soon after our release from Hussain Kamel's internment, Jafar asked for me and enquired whether I had any magneto optical discs left. He knew me too well for not failing to have kept at least one redundant disc for emergency situations. I did have one, the disc that was left as a gift by the Japanese Canon representative, who sold us the first two devices, and was hence not accounted for in the proper inventory. Jafar was apprehensive of David Kay's aggressive methods. He had many documents in his personal library at his house. The information was highly sensitive. Jafar sent his driver, the ever-faithful Omran, my companion in the prospecting for uranium, to my home, where I had stored one of the CanoFiles, with a box load of sensitive reports, documents and papers. Omran was not to lose sight, for one minute, of the contents of that box. I set up the CanoFile on the dining table and Omran started handing me the documents as I indexed and stored them. I did not allow even my mother to enter the dinning room except to hand us tea at the door. I glimpsed on the monitor, whenever I had the chance, to see what was being archived and skimmed their contents. I was not that pleased with some of it. It was plainly a blatant exaggeration, or promising extrapolation, of what we were achieving using the Baghdatrons in 1990, signed by Jafar and presented by Hussain Kamel to Saddam. It would have been misleading as to the true progress of the work, if you do not grasp scientific innuendo. In the event, Omran left with a filled magneto optical disk and his box of documents after five hours of work.

The Foreign Ministry

Several events slowly tore me apart from the Centre for Specialized Information, though its service and value to the rehabilitation work was prospering and its benefits appreciated by many. At the end of the summer of 1991, I was paid a visit by the Head of PC3 Security who sternly informed

The Foreign Minister, Mohammad Said al-Sahhaf, pinning a medal, 1996.

me that I should hand over the two CanoFiles to his agents. I strained to find out the reasons for this order vainly pointing out two main problems. First, we had used most of the five discs that we had originally bought to store PC3's documents. Secondly, proper archiving depended heavily upon a rigid procedure for indexing the documents that were to be saved; otherwise, it will be very difficult retrieving what was stored. He adamantly refused to tell me the reason for their confiscation. He grudgingly accepted my offer to have Salam Toma train his security agents on the use of the instruments but would not allow Salam to show them how to index the actual documents themselves lest he took a glimpse at what was being archived. He was utterly overconfident in the ability of his officers to master the indexing process. He would see to it that more discs would be provided. Reluctantly, I handed over the two machines fully convinced that this would turn out to be a failed effort.

Adil Fiadh, still following our rigid procurement procedures, forwarded me a quotation for the purchase of fifty
CanoFile discs to be procured from Jordan, for my approval.
I was shocked at the price, which was more than double
what it should have been. I refused the purchase. I was even
more shocked when Adil Fiadh himself paid me a visit to my
office and in hushed tones implored me to rescind my decision, hinting to very high authorities. The point sank in.
Hussain Kamel was behind all of this, notwithstanding the
proceeds of the kickback from such exuberant price. This
fiasco was in sharp contrast to the absence, as far as I am
concerned, of known kickbacks throughout my previous 25
years of work in the Iraqi Atomic Energy Commission. The
economic deep pit holes portending the difficult years ahead
were forming.

Several months later, the Head of PC3 Security returned
and asked for Salam Toma's help. His agents could not
retrieve what they had stored. Salam found the two instruments in a store among sacks of flour and rice. They have
been rendered close to useless. They were barely functioning
as the security agents carelessly fed in documents with the
metal staples still embedded in them. As expected, Salam
found out that they simply fed in the documents to be
scanned without any formal indexing. The documents
belonged to Khalid's Group 4. They were mainly administrative papers. It was part of Hussain Kamel's attempt to
hide reports and documents concerning the nuclear, biological and chemical weapons programs in his now famous
"chicken farm". However, the value of these particular documents was largely useless. I demanded and received the two
failed instruments. They had only used two out of the fifty
purchased discs. The other discs disappeared, only to reap-
pear several years later when I re-purchased them locally.
The fact that I had these CanoFiles opened the door to my
close relationship with Mohammad al-Sahhaf, our Foreign

Minister at the time, and the famed Information Minister during the 2003 war. After the occupation, he remained hidden until he appeared on Arabic news channel al-Arabiya and Abu Dhabi TV on June 26, 2003. Speaking to al-Arabiya, al-Sahhaf claimed that he had made contact with the US forces through friends, was questioned and later released. Al-Sahhaf was not on the US list of 55 most wanted Iraqis, the so-called pack of cards. Al-Sahhaf finally arrived to Dubai with his family in July 2003 and intends to complete his memoirs.

Al-Sahhaf is a self-educated gentleman with a rigorous personality. Though the dictates of high office in the Iraqi government demanded long working hours, al-Sahhaf's workaholic affliction was legendary. Always susceptible to a good joke accompanied by a hearty laugh, he treated his colleagues with compassion and fairness, yet firm and strictly abiding by the rules. I continue to cherish his support and brotherly concern for the harsh times that then lay ahead.

Two particular incidents finally tore me from the Centre for Specialized Information and into al-Sahhaf's welcoming arms. The first was a reprimand for the work rendered to Humam Abdul Khaliq, the previous head of PC3. After the 1991 war, and the collapse of PC3, Humam was appointed the Minister of Higher Education and Scientific Research. He later became the Information Minister, to be superseded in 2001 by al-Sahhaf, while he returned to the Ministry of Higher Education and Scientific Research. Humam was captured by the Americans in late April 2003, and is rumored to have been released in May 2003.

Humam had learned in 1992 that I had embarked on a self-initiated campaign to network the MIC research centers via dial-up modems with a central CD-ROM databases collection at the Ministry of Industry. He called me over and requested that I would also consider connecting all of the universities in Iraq within this nascent network. I obliged and within a year had handed him a report and a diagram of the about 65 universities, research centers and ministries that Ayad Muhaimid and I had managed to network within the span of two years. He forwarded that report with its diagram to Amir al-Ubaidi, the Head of the MIC recommending a citation and a generous bonus for such an accomplishment. Instead, Amir handed me yet another heavy scolding, a reprimand and a cut in salary for what he claimed to be a "security breach" by exposing the names of MIC's research centers on that diagram.

The second incident was the heavy-handed intrusion by MIC's Security into the workings of the Centre for Specialized Information, which was by 1993 an integral part of MIC along with the rest of PC3's departments, as PC3 itself had ceased to exist by then. The loose and responsibility oriented managerial style with which I ran the Centre did not conform to the strict military-like security procedures of MIC. A trainee, who claimed that he would like to assist us, joined our small group. Upon a tip from Salam, I later deduced that he was planted. I set a small test trap for him and he fell for it. He had written many reports to Walid, the Head of MIC Security on the daily ongoing activities at the Centre. I fired him immediately, but it was too late. Amer al-Ubaidi, the Head of MIC, had ordered an official investigation into the workings of the Centre by a five-member high-level MIC committee. The accusations consisted of irregularities in the paid salaries, loose administrative controls, and unauthorized travel across Iraq in my government car as I established the information retrieval network as well as having some of my female employees dance in my office. That last accusation was a colorful exaggeration of having played an audio CD of Um Kalthoum, the most famous and venerated of Arab singers, on my up-to-date personal computer in my office, which was a great novelty at that time, and inviting my staff, including the devout Moslem Nahida, to come

listen to her glorious voice. All of these accusations were disproved and fell by the side. Yet, Walid insisted on a closure
on a technicality that would save his face. He insisted upon
holding a final meeting of the MIC committee, which I
refused to grant. A Christian member of the team called me
imploring my consent to close the affair. I took a decision
and agreed. When they finally assembled in my office, I started talking in the Christian dialect with the committee member who had convinced me, to his utter astonishment at this
unaccustomed breach of formal procedure. I had a point to
make though. I took a blank piece of paper, signed and dated
it and handed it to Walid. He was taken aback asking for the
meaning of this. I explained that I authorized him to write
whatever he liked about my character, that I am religiously
biased (an anathema in Iraqi official circles), a gay, a
debaucher, a liar and whatever his imagination may fancy,
but not to transgress on my loyalty to my work or to my
country. Would he then do me a favor and use whatever he
accuses me of to kick me out of MIC? With that, I left the
meeting and headed for al-Sahhaf.

Al-Sahhaf had already invited me to his ministry to show
him the workings of the CanoFile, upon the recommendation of Abid Illah al-Daiwachi, a mercurial colleague who
was the perennial Head of the Computer Centre at the
Ministry of Industry. Al-Sahhaf had an inherent and lasting
wonderment at the workings of the personal computer and
the benefits of the various softwares. He had proposed that I
would assist him in introducing the use of personal computers to the Foreign Ministry with the ultimate aim of networking its departments, managing its voluminous daily
documentation electronically and to archive its rich historical files. He took it upon himself, being a close friend of
Amir al-Ubaidi the head of MIC, to convince Amir to allow
me to visit the Foreign Ministry a few days a week to start
on this enterprise. Amir agreed, but only verbally, for he

would not dare make the matter official in writing and titillate the attention of the highest MIC authority, Hussain Kamel, who was not on close terms with al-Sahhaf. In the light of the two above events with MIC, I flagrantly extended the approved few days a week to cover all days of the week, despite Amir's complaints and under al-Sahhaf's protection.

From the start, I explained to al-Sahhaf my intentions to secure the education of my children abroad. He was very understanding and considerate. He promised to assist me to the best of his ability, not by releasing me from the civil service but by securing a diplomatic post for me, in New York for example, working with my life long friend and also a high school colleague, Nizar Hamdoon. Nizar was Iraq's representative to the UN at the time. He regretfully passed away on July 4, 2003 succumbing to a terminal ailment. His political "will" is thoroughly Iraqi in its mature approach to extricate Iraq from its present free fall[25].

I accepted al-Sahhaf's promise while realizing his limitation. In order for him to thus assist me, we would have had to go through Hussain Kamel and have him agree to my transfer to the Foreign Ministry, a daunting and formidable task, to say the very least, as the two of them disliked each other immensely.

Al-Sahhaf financed the repair of the two CanoFiles and bought a third one from Jordan. We started using them in archiving the Foreign Ministry's extensive historical records. Abdul Halim al-Hajjaj, the elusive shadow of Khalid Said, paid us a surprise visit in 1994. He was at the time deputy-in-charge of the Iraqi Atomic Energy Commission and coordinating its negotiations with the IAEA inspectors. He had brought with him Group 4's magneto optical disc. It was one

---

[25] "It's Not Over Until Saddam Is Over", Washington Post, Sunday, July 27, 2003; Page B04
http://www.washingtonpost.com/wp-dyn/articles/A48705-2003Jul25.html

of three discs, the other two contained the images of the rest of PC3's reports, that I had kept separate from the cache that was found by David Kay in September 1991, and had so strenuously objected to hand over to him but for the dire command of Jafar.

Abdul Halim requested to have a couple of copies to be made of it. At the time, I was surprised at the ease with which he had managed to penetrate the security layers of the Foreign Ministry and simply show up at our Computer Center, with permissions already granted for his request. In any case, since the limited number of discs we had at the time was being used for archiving the Ministry's records, we had to obtain some new ones. I spread the word around, and lo and behold, a whole stack of them surfaced. They were the remnants of the fifty discs that were exuberantly bought by Adil Fiadh in 1991 but then disappeared when Hussain Kamel's security staff had fumbled their use and had damaged the CanoFiles. These discs were bought by Abdul Halim al-Hajjaj at much reduced price with no questions asked.

After providing him with the requested copies, Abdul Halim had another request, which was immediately approved by al-Sahhaf. Perhaps I would accompany him to Tuwaitha where he was to meet the representatives of the IAEA in order that I could hand over to them this copied disc and respond to any further technical questions that they might have on the CanoFile. Apparently, they were only concerned about Group 4's activities that entailed the design of the bomb, and did not ask for, or showed interest, on the uranium enrichment activities of PC3. The copied magneto optical disc was duly handed over to them. I am not sure whether they knew of, or even now have, the other two discs.

Upon my constant and persistent imploration to al-Sahhaf to attempt to release me from MIC and transfer me officially to the Foreign Ministry so that I would perhaps be able to go to New York, he relented and in turn implored and convinced his friend Amer al-Ubaidi to gather enough courage to confront Hussain Kamel with this request. Hussain Kamel was more than furious when he learned of my whereabouts for the last two years and delivered a bruising kick to the shin. I was ordered to immediately return to the Centre for Specialized Information, which was fully integrated with MIC's information centre by then and under their direction. I would loose much seniority and many of the privileges that I had garnered.

Stunned by this sudden turn of events, I scurried around to find a better solace. Abid Illah Dewachi, who had introduced me to al-Sahhaf, was by then assigned by Hussain Kamel to setup and head a company that looked after the obligations of former computer companies, such as HP, IBM and NCR. I applied to join his organization. He sensed the effort and in turn attempted, for reasons known to himself, to block my request. I turned to a higher authority than Abid Illah. I sought the assistance of Khalid Said. He was a top administrator at the time in MIC. Khalid adroitly managed to stem Abid Illah's rush to stop me from joining his organization and skillfully thrust me upon him. In late 1995, I took an office in Abid Illah's new building and kept my government car and most of my other privileges.

Once settled in, I rebelled, for the third time during my government service. I refused to work. I would go to work daily, sit behind my table and not turn on the lights until it was departure time. This lasted for eight months. Abid Illah desperately tried to convince me otherwise. I suggested to him that the only solution would be my retirement. After six months of my self-imposed internment, Abid Illah finally gave up on me, and took it upon himself to convince Amir al-Ubaidi that I was a dead piece of wood and might as well try to convince Hussain Kamel to throw me out. Amir, already fed up with my case, made one final attempt. However, Hussain Kamel threw my file in his face when

Amer proposed to him that I should all together retire. Amir angrily demanded from Abid Illah to inform me never to talk to him again or ask him for anything.

Two months later, I was with Abid Illah in his car when he mentioned a strange non-event. It was the 17th of July festivities for the Ba'ath party and Hussain Kamel had missed the grand celebration of the event at MIC.

Rumors had it that he had escaped to Jordan.

Immediately, I called al-Sahhaf. He instructed me to visit Amer al-Ubaidi and to get an official transfer to the Foreign Ministry. Amer al-Ubaidi, who did thankfully allow me to talk to him again, had already been removed from MIC and assigned to be the Oil Minister. Nevertheless, he issued a backdated order for my official transfer. Al-Sahhaf was more than pleased and immediately secured for me a diplomatic passport, as a mid-level diplomat. It was September 1995 and I had just missed the Iraqi delegation going to the UN General Assembly's annual meeting. Al-Sahhaf promised that I would definitely be part of the delegation going to New York the following year.

Within a period of three years, and with al-Sahhaf's generous support and determined drive, we had laid out a computer network throughout the Foreign Ministry's departments, installed a document scanning and retrieval application that allowed al-Sahhaf, from his office chair, to recall an image of a document that went to or from any of his departments. The Computer Center occupied the Ministry's former posh dining room and the total staff was about five. Every several weeks, al-Sahhaf would drop by and ask me for a list of required software and hardware to bring back with him during his frequent travels abroad. He had subscribed to several computer weekly magazines that, to my surprise, he would have read or at least glimpsed through, before handing them to me. From them, I would choose and make a list of the latest software and storage hardware that I felt was needed for the computer center, or just useful to have. I would unerringly receive the contents of the whole list upon al-Sahhaf's return, as he went himself to the computer stores to buy them, enjoying his personal computer hobby.

We also managed to train about 250 diplomats on the use of word processing. That is when I lost a pound of flesh due to a rabid Intelligence officer who left a life threatening mark on my record for several years hence.

A national election to select the members of the national parliament was about to take place at the end of 1995. In order to prepare computerized lists of the eligible voters, each ministry in Baghdad was assigned a few geographical districts. Hand written lists, prepared by the Ba'ath party locals in these districts, were submitted to each ministry to be converted to a computerized and alphabetically sorted listing. Faced with about 40,000 names to be typed, I suggested to al-Sahhaf that this would be a perfect opportunity to have the recent batch of diplomats, who have freshly graduated from our word processing course, to practice what they have been trained on and to complete this task at the same time. He immediately concurred and issued the order for the 80 diplomats to receive their floppy discs, instructions and 500 names each from the Computer Centre.

Most of them obliged. When al-Sahhaf asked me a few days later on the progress of the work, I mentioned that 75 diplomats did receive their assignment but five had not. Who were they, he angrily enquired. I did not know them well, but they were in the Expatriate Department and I recited their names. "The Intelligence bastards" he exclaimed picking up the phone. Realizing the danger of the predicament that I had created for myself, I implored him to calm down for fear of "what bites these dogs might inflict upon me". He was beyond reason and assured me that he would defend me. He talked to the head of that department in scolding terms, ordered that these five Intelligence officers would better go down, and receive their floppy discs and the list of names within half an hour or he will kick them in the butt to back where they came from.

I could plainly see the poison in the eyes of their leader, Salah Abdul Rahman al-Hadithi (whom I later learned was in charge of the US section of Intelligence) when they filed by and received their typing quota. He indignantly claimed that they had enough typists in Intelligence to do this work. I retorted with a quote from Saddam that curtly shut him up. Yet the damage was already done, and intensively.

Only three years later, did I find out that he had immediately filed a report to Saddam pointing out that my name was mentioned in a New York Times article as having a possible connection with the Israeli Mossad and that this was being secretively investigated by his diligent efforts. He also referred to al-Sahhaf's attempt to secure a diplomatic post for me abroad. Saddam had scribbled that I "shall not see the Iraqi border" in my life, in red ink. That paper, stored in my Intelligence file and unknown to al-Sahhaf, had sealed my fate for good, as far as any attempt to travel abroad was concerned, and no matter whom in the Iraqi government might try to get me out. It was a Saddam red ink unchallengeable dictum.

This was Salah al-Hadithi's devilish way of getting back at both al-Sahhaf and myself, with one blow.

During 1996, Iraq was embarked on detailed negotiations with the UN on the Memorandum of Understanding (MOU) or the Oil for Food program. These strenuous negotiations were held at the Foreign Ministry. By virtue of my position as Head of the Computer Center, where all draft agreements ended up to be word processed in English and Arabic, I started to attend the actual negotiation sessions to implement more efficiently the preparation of the final document. It was at these meetings that I came in contact with some of the heads of the UN agencies in Baghdad. I dreamt of retiring from the government and working for them as they paid 50 times the government salary, as well as being a springboard for further work abroad. In the spring of 1996, a position had come up at the United Nations Development Program (UNDP) in Baghdad. I called my friend, Shirin al-Jaff who worked there, enquiring whether my wife, Niran, would fit the position. She boldly proposed that I myself should apply, as the job fits my qualifications perfectly. I told al-Sahhaf of this opportunity and requested his permission to apply. He was encouraging, even though I was still worked for him. In order to allow me to apply, he requested his senior deputy minister, Saad al-Faisal, in charge of the Ministry's Intelligence, to arrange for the Intelligence's permission to work with a foreign institution (Saad was captured by the Americans in late May, 2003). A meeting with an Intelligence officer was held in Saad's office during which a short interview culminated in a stamp of approval by the Intelligence. No favor was asked in return.

I, along with 250 others holding PhDs and Master of Science degrees, applied for that one job. Such were the dire economic circumstances in Iraq in the mid-nineties. The list was whittled down to 18 who were then given written and oral tests. I ended up in the three final contenders who were interviewed separately by the heads of the UN agencies. A friendly phone call from abroad landed a friend of mine the job, though he did not do well at all in his written test. However, my interview with the heads of the UN agencies had apparently left with some of them a few lingering impressions about my skills and experiences. This proved to be decisive a few months later, as events hurriedly unfolded.

In the summer of 1996 and armed with my diplomatic passport, al-Sahhaf had secured for me a presidential order to accompany the highly competent Riyadh al-Qaisi, the deputy minister, for an extended trip, starting in mid-August 1996, to New York, Paris, Moscow and finally Cairo. We prepared all the paper work, required visas, and plane tickets and obtained our foreign currency stipend quota. One day before our departure, and as Riyadh and I huddled with the Russian ambassador for final preparatory discussions for the trip, Riyadh instructed his driver to obtain the exit visas for both of us. An exit visa was required for all Iraqis who intended to travel abroad. Ordinary civilians needed to pay 400,000 Iraqi Dinars for it. The driver returned after one hour announcing that they only gave Riyadh his exit visa, but objected to mine. The stated reason was that I had a four-part name, but no tribal name.

A digression into an Arabic semantic intricacy is in order at this point.

It is an Arabic custom to call a man/woman by the name of his/her son. Abu would stand for a father and Um for a mother. The reason being, one would assume, is that when you would call a man/woman by that paternal extension to the name, you would immediately appeal to their warm instincts and smoothen the listener to whatever follows. The designation for the son's name is often historical and is passed on for generations. For example, Tariq Ibin Ziad and Khalid Ibin Walid were famous Arabic military generals. Ibin means son in Arabic. In the abundant richness and long lasting flourish of the Arabic language, the Arabs maintained their respect for these leaders by preserving their full names ad infinitum. Hence, if you are called Ziad, then your father should be Tariq, such as the case with Tariq Aziz (the famed Iraqi diplomat), whose eldest son is called Ziad, taking after the legendary Tariq Ibin Ziad. You would then refer to Tariq Aziz, even officially, as Abu Ziad. My brother Walid named his son Khalid, so Walid is addressed as Abu Khalid, after the legendary Khalid Ibin Walid. Even if you are not married and do not have a son, people would still start calling you Abu (whatever your son's name should be) according to your own name. This is done out of a sign of respect once having passed the puberty stage. Hence, your son's name is predetermined by tradition. Or, if your name does not fit in one of these established historical names, you would still be called Abu Ghaib, meaning the father of the absent one, waiting for you to have and name your eldest son, so that you may henceforth be called as Abu your eldest son's name, and perhaps setting a new linguistic tradition if you do become famous. This became so widespread and common, that we would sometimes be at loss to know the full name of a friend, since he was always, and for years, been simply called Abu Something.

My father claims that our family's ancestry descended from a Yazidi clan. This religious sect still flourishes in the north of Iraq. They worship both God and the devil. However, they claim that God is already good and benevolent and hence you may simply accept his grace. In contrast, you should really appease the devil in order to ward off his wrath and mischief. Relying on having seen some old documents, my father claimed that the Khadduri Yazidi family had some sort of a royal standing about three hundred years ago but were kicked from the Yazidi clan for some reasons, now unknown. Part of that extended family had traveled to Basra, the main city in southern Iraq, and the other part had remained in Mosul, the main city in the north. The renowned historian, Majid Khadduri, who was the PhD supervisor of my brother Walid at Johns Hopkins University, belongs to this northern branch. We are the Christian ancestors of the Basra branch. The Arabic naming tradition had also been applied to our family. In the Arabic Iraqi Christian naming tradition, Yaqoub is the father of Khadduri. (Interestingly, the same name is Kaddoori for a Moslem and Khadhouri for a Jew, hence easily distinguishable as far as the religion of a family is concerned). Therefore, my full name would historically be Imad Yousif Yaqoub Khadduri Yaqoub Khadduri Yaqoub Khadduri. My grand father, Yaqoub, had, for reasons of his own, broken that tradition and named his eldest son, my father, Yousif instead of the traditionally ascribed Khadduri. Having regretted that decision, probably due to peer pressure and chastisement, he relented and named his second son, my beloved uncle, Khadduri. Alas, the die was cast.

Most of the people in Iraq belong to one of the many Arab tribes. Tribal law, which is parallel, and in certain circumstances more dominant than civil law, is a potent force in Iraqi society. Some families do not belong outright to a tribe, as was the case for my family. In my official records, my name is simply Imad Yousif Yaqoub Khadduri. However, having obtained one of the official distinctions that did require a tribal name, I was conveniently bestowed the tribal name of Khadduri on one official card. My name, as on that card, was Imad Yousif Yaqoub Khadduri Khadduri.

Hence, with a rejected exit visa, I was ready to go back with the driver to prove my tribal name as shown on the card that I had on me when Riyadh al-Qaisi, in his wider experience, calmed me down and announced, to my initial surprise, that this is not a name matter, but rather a matter for the minister to resolve. He adjourned the meeting with the Russian ambassador and headed for al-Sahhaf, who immediately left the Ministry for the Presidential Palace.

Later that day, al-Sahhaf explained to me the stated reason for the rejected exit visa. He claimed that the recent Iraqi military push into northern Arbil to oust Ahmad al-Chalabi's CIA-backed supporters had caused heightened security measures. Due to my sensitive background, I could not at that moment in time be granted an exit visa. He promised to attach me, however, to his own entourage that was heading for the UN in a month's time towards the end of September 1996; by that time, matters should have cooled down. Al-Sahhaf was, I still believe, not yet cognizant of the existence of Salah al-Hadithi's poisonous report lurking in my secret Intelligence file.

When his trip became due, and after persistent attempts, he finally called me on the last night before his departure. He had, yet again, failed to get me an exit visa. I was expecting this. "We have met on cordial terms", I stated, "may we part on similar terms" and I handed him my retirement request. He sadly approved it.

Salah al-Hadithi's poison had exacted its first solid revenge. It would not be the last time. In the words of my rescuer, three years later, "with that report in your Intelligence file, you should have never, even in your dreams, been able to reach the Iraqi border".

The next day, and after returning my short lived and unused diplomatic passport, I held in my hand the long sought after letter of retirement signed by Tariq Aziz, who was in charge of the Foreign Ministry in the absence of al-Sahhaf. I headed to Habib Rejeb, the Head of the World Health Organization (WHO) agency in Baghdad who had expressed interest in my services on the condition of my retirement. He kept his word, and on the next day, I started work at the WHO. My assignment was to again network, via telephone, the various Ministry of Health warehouses that are spread throughout Iraq in order to maintain proper monitoring and control of the inventory of medicines and medical supplies that were being acquired through the Oil for Food program, under the supervision of WHO, and was intended to be distributed to these warehouses.

In addition, having the letter of retirement meant that I was to start the countdown of the two years probation that was imposed a year earlier by the government on all employees of the Atomic Energy Commission, including the former PC3 staff. No one could obtain a passport or travel abroad until the end of that two-year probation starting from the date of his retirement.

Though stubbornly hoping to be thus able to leave Iraq

legally, Niran and I had already warily treaded the path of escaping from Iraq.

The escape

Within a few years after the end of the 1991 war, Niran's colleagues at the private al-Mansoor University College, where she was teaching computer languages, were slowly slipping out of Iraq, some legally while others illegally, one of them being Khidir Hamza who lectured there. Since there was no air travel due to sanctions throughout the nineties, the escape routes were limited through the risky north to Turkey or the dangerous border point of Traibil to Jordan. Niran would calmly break the news to me in her sustained slow persuasive manner. When her close friend, Samira Katoola and her husband Tawfiq, both former Atomic Energy Commission PhDs, also suddenly vanished in 1994, along with their children, after a well hushed-up departure, we decided to find out their route and try it ourselves.

News, whether true or gossip, and except for very rare Intelligence and Security snippets, is hard to hide in Baghdad. We soon learned that a certain Abu Abdallah in the northern city of Mosul was the conduit through which that family had slipped out. Further contacts through my cousin, Salam Khadduri, who got me in touch with Kurdish smugglers, confirmed the news. Furthermore, Abu Abdallah was a Christian; hence, he was more trustworthy in our eyes.

Niran and I had several problems to surmount as we groped for a way for our family to leave. The first obstacle was that she was still 43 years old and the Iraqi law stipulated that no woman under the age of 45 might travel abroad without the accompaniment of her husband, father, brother or uncle. The reason for that was ostensibly to lessen the opportunity for Iraqi prostitutes to travel abroad. She also held a Masters of Science degree in Computer Science that hindered her attempt to obtain an exit visa as the directives clamped down on holders of higher education degrees to prevent their departure from Iraq. The other obstacle was that her name, as well as those of my children, was coupled to my name in the Intelligence and Security databases, especially at the Security passport offices. Therefore, any attempt to obtain an exit visa for any of them would ring bells at the Intelligence Agency that members of my family were planning to travel, a taboo without the Agency's explicit permission.

Semi-convinced on the futility of this escape approach, Niran and I, nevertheless, traveled to Mosul to meet Abu Abdallah and his very capable hairdresser wife. I had the passports for my family that I had obtained a week after the end of the 1991 war, but did not have a passport for myself. Abu Abdallah had all the tools of the trade in his living room; photocopier, white ink erasers, pens, and empty request forms for an exit visa. To prove his capabilities to us, he immediately sat down with our passports and completed the application forms for the exit visas for Niran and the children. He assured us that neither Niran's age limit, nor her degree, would be a hindrance. Furthermore, the Security database in Baghdad will not be informed of the issuance of the exit visas for the members of my family. He confidently walked us with him into the Mosul passport office, which was a few steps from his home, to the warm greetings of several key Security officers who relished the benevolent extra income gleaned through him. He swiftly acquired the final few official stamps and signatures on the forms in his hands. Abu Abdallah and his wife offered us the following scenario. They would guarantee the arrival of Niran and the children to Beirut, Lebanon, via Turkey then Syria. It would then be up to us to appeal there through Christian societies (they provided the names of some of the priests there) to obtain immigration visas or, if we paid enough, even apply for Lebanese citizenship. He claimed that Samira and Tawfiq had made it via that route to Cyprus, where they were awaiting their papers to go to the UK.

Abu Abdallah's influence and muscle stopped at the boundary of the Mosul passport office. He would not even contemplate issuing me a fake passport with my sensitive connection to the Iraqi Atomic Energy Commission, which was embedded in so many Security and Intelligence databases. I would have to fend for myself for that.

We partook lunch with them. All that remained to get the passports stamped with the exit visas was the matter of handing over one thousand dollars. Niran and I excused ourselves to sleep the noon siesta at the hotel and to make our decision. I was sitting on the hotel room balcony staring blandly over the wide Mosul horizon. Niran stepped into the balcony and I looked at her. We shook our heads, hugged and returned to Baghdad.

Two years later, we were invited to a social gathering at Niran's sister, Nariman and her husband Saad Yono. They were giving a goodbye party to a mutual friend, Khalid Rumaia, a senior mechanical engineer at PC3. I was piqued. How did he manage to get his passport? He took me aside and whispered the tantalizing details. First, he had managed to retire, citing medical problems, a couple of years earlier. Then he had managed to get, in confidence, to Khalid Rashid, the administrative head of the Security section at MIC. Upon the integration of PC3 with MIC, all PC3 staff had fallen under his security jurisdiction. A mere $400 was all that it took for Khalid Rashid to slip Khalid Rumaia through a loophole. MIC's security office was more than closely linked with both Internal Security that issued the passports and the Intelligence Agency. Not to be inundated with unnecessary paperwork, the Intelligence Agency set out a guideline for Khalid Rashid obligating him to forward to them only those valid requests (those that spent the two years probationary period after their retirement) of very senior staff, those that were at a level of General Directors and above, for their own scrutiny and approval. All other valid requests from lower level PC3 civil servants were left to his discretion and according to their sensitivity as documented in their files. That $400 made sure that Khalid Rumaia's sensitivity was so low that an approval by Khalid Rashid to the Internal Security was enough to secure Khalid Rumaia and his family their passports and exit visas.

Khalid Rumaia bid us farewell after leaving me Khalid Rashid's private phone number, with the advice to make sure to mention that it was he who had given me that number when I did make the call. By then, I had just retired, working for WHO and counting the days for the passage of the two years probationary period. My eldest kids still had a couple of years ahead of them before finishing high school. I made sure that I did not loose that number.

Right next door was Niran's other sister, Noorhan and her husband Amer Saman who had a Masters of Science in Statistics and worked on rocket designs at one of MIC's military institutions. His elder brother was already a US citizen and had forged ahead years earlier with the required paper work for Amir's immigration to the US, along with his family. The main problem for Amer was how to leave Iraq, as he was banned from getting a passport because of his sensitive position. Furthermore, Noorhan was still 40. To complicate matters a bit, he did not, in my opinion, have the courageous stamina required for such a challenging undertaking.

By virtue of neighborly ties, Noorhan and Amir had learned that a certain Christian, Bassim Ysho' Putrus, henceforth referred to by his traditional name, Abu Diyar, had managed to smuggle out their across-the-street neighbor effortlessly to Jordan. They contacted him. By his amazing confidence and calm resolve, Abu Diyar managed to grab Amer by the neck of his shirt, issued him a passport posing him as the uncle of his own wife in order to accompany her outside Iraq.

He issued him another valid passport in Amer's true name, which was useless in Iraq but was, nevertheless, duly signed by the Security officials as he passed the Iraqi border, and with which he entered Jordan. He stayed with Amer as he bartered with the American Embassy to convince them of his value and authenticity. Abu Diyar returned to Baghdad when Amir's family finally left to the US. His fee was $1000.

That was proof enough for Niran and me.

The ready passports of Niran and the children were a cinch for Abu Diyar to ready with exit visas. My sensitive position was much more difficult than Amer's. After initial enquiries, Abu Diyar had found out that my name is cross-referenced in so many Security and Intelligence databases that he could not, with all of his reach and contacts, issue me an "authentic" passport with my true name on it. The only way to get that passport was through the Military Industrialization Corporation. Abu Diyar offered to issue me a passport with a different name. By doing that, he would be able to get all of us out of Iraq smoothly and together.

I was not ready for that alternative. I had always insisted on leaving Iraq legally. I had nothing to hide or felt ashamed of and did not see any need to be entangled in new identities. I had already retired and was counting my two years probationary period before being allowed to apply for an authentic passport. The children still had a year or two to grow into before university. We would wait it out, though Niran and I had already been secretly working on our intended departure. Abu Diyar respected our wishes and stood by to offer any required assistance, even though the border gates were being slowly shut by the day, and ever so tightly.

Upon Niran's constant and persuasive urgings to get some paper work moving I finally completed the single page form to apply for a Landed Immigrant application to

Canada. The filled form was secretly sent in 1995 to the Canadian Consulate in Amman, Jordan through the auspices of Subhi Ayyoub, a most trusted and reliable Jordanian friend. We waited for their response for about a year, and despite two reminders, not an indication came back. This was no mere waiting, but tensed nerves and flailing hope as Subhi would travel to Amman and return to Baghdad every couple of months, empty handed.

Early in 1996, as we chatted across the fence with our life long neighbor, Luma al-Saigh, on her experience in gaining a Canadian Landed Immigration foothold, she innocently asked us about the number of points that I had qualified for. Points, what points? Any conversations about such topics and to such places would always take place in the utmost secrecy and in the lowest volume, which usually meant a lack of full details. She went into her house and brought out the information sheet on how to calculate the points that we apparently had not realized the need to sum up. Luma surreptitiously handed it to us over the fence. With zero points for my age being over 50 and the worth of a mere 2 points for 25 years experience with a PhD in nuclear reactor technology (while a certificate in French cuisine garnered 23 points), it was quite obvious that my total of 48 points was no where near the 75 points threshold, and hence the utter lack of the Canadian response. Dejected at losing a whole year just to find out that piece of information, Niran, in her calm and steadfast manner, sat down and calculated her own points. She had six months left before reaching the age of 45 when she would lose 2 points. With her MSc in Computer Science from the UK, she had accumulated 76 points. Subhi dependably delivered Niran's application and when he returned from Amman two weeks later, he handed us the Canadian four-page application form for Landed Immigrants.

That Masters degree in Computer Science, that I had so

insisted that Niran should acquire when we got married in 1976, had paved the way for us and guaranteed our family the beginning of a new fork in life.

Subhi himself was taking great risks throughout this period in delivering our mail and returning the Canadian responses, as any thorough chance search at the border would have cost him, and us, immensely.

From the outset, we had made it very clear to the Canadians through our application that my particular situation and background were very sensitive. Furthermore, Niran was teaching at a private college and they had just enacted new regulations barring from travel abroad any teaching staff with higher degrees, in the public as well as the private universities and colleges. Hence, we pleaded for a pre-approval or at least the assurance of such approval from the Canadians so that we may fully depend on it and plan a one time risky exit. We would need to stay as short a time in Jordan as possible in order to avoid risking being caught up and dealt with by the Iraqi Intelligence agents in Amman. The most the Canadians would offer us was that Niran must be interviewed in Amman, but that I may be interviewed in any other capital of the world. The other impression that we got from their response was that it would suffice for them to have only Niran for the interview, and not both of us. We commenced on this presumption to painfully plan accordingly, to our dismay later.

For the moment, in the spring of 1997, I was approaching the end of my two years probationary period after retiring from government service. I had spent six months at WHO, installed the health network across Iraq and trained more than 60 pharmacists and warehouse supervisors on the use of the network and the database. I was replaced, at a whim, by a junior Egyptian computer specialist, barely 21 years old. His uncle was a consultant at WHO and had paid a work visit to Habib Rejeb, the head of WHO in Baghdad.

He simply exercised the all too well known comradely spirit prevalent throughout the UN organizations: you rub my back and I will rub yours. I, the Iraqi, lost my job with no explanation offered.

I was very fortunate that Shirin al-Jaff, who had first enticed me to work for the UN, was still looking for a qualified National Observer to assist the United Nations Development Program (UNDP) in Iraq to implement its responsibility for the rehabilitation of the electric sector in central and southern Iraq and to coordinate its implementation in the Kurdish north of Iraq under the Oil for Food program. I had an extensive network of contacts, scientific background, and first hand experience with the electric rehabilitation effort in Iraq. Furthermore, the success of organizing the three electric power rehabilitation symposiums in the summer of 1991 resonated well with UNDP procedures and tempo. I joined UNDP a month after leaving WHO in May 1997.

I called Khalid Rashid at MIC on his private phone number and confined to him that Khalid Rumaia, who had by then settled in the US, sends him his greetings. He graciously accepted the gesture. We met and agreed that my fee would be $600, and not $400 as for Khalid Rumaia as it was a bit riskier for him to downplay my importance. He was tall, lanky and handsome. Grasping the opportunity of my work at UNDP, he interceded on behalf of his girl friend who was looking for work there. I tried to put as much a responsive face to his request as feasible without stepping on anybody's toes. His impassioned telephone pleadings forced even the novice in me to warn him that the phones were not safe and might expose us. I finally relented and invited her for an interview. He did have an exquisite taste in women. Nevertheless, I had to stretch the gesture till I got my passport. I did pay him many evening visits to his lowly dwellings with his quarreling wife, stretching my promises and stacking up on beer before each encounter.

As the end of the two years of probation approached, Khalid Rashid prepared a confidential enquiring letter to my superior, Jafar Dhia Jafar, on whether, in his opinion, I still harbored any nuclear weapons program secrets after the two years of retirement. He had it delivered by hand to Jafar's office. Upon Khalid's recommendation, I had to closely follow the letter's path at every turn and station, inconspicuously hinting to my friends at Jafar's office who handled it to not mention what they · were passing around lest the Intelligence Agency got a whiff of it. Even though Khalid's own authority extended up to people like me, but not to a higher administrative position, its exposure would still potentially nullify his attempt at utilizing this loophole if we are not careful in concealing the attempt from the omnipresent Intelligence.

Jafar, realizing the hush-hush tone of the matter, procrastinated a week or two since we were in the midst of preparing our final declaration on the whole scope of the nuclear weapons program, before and after the 1991 war. We were about to submit the total gamut of our activities in that report to the IAEA. Many meetings were held with all of our colleagues in the now defunct PC3, some of whom I had not seen for several years. We would spend countless hours in the secret house in Jadiriyah district going over reports, papers and faint memories. One sticking point was the fate of the magneto optical discs that Jafar had made me hand over to Abdul Halim al-Hajjaj before the start of the 1991 war. The disappearance of the two discs that contained Groups 1, 2 and 3 reports angered Jafar, who did not clearly recall my strenuous objections to letting go of them in the first place. Jafar finally convened a meeting with all remaining staff of the disbanded documentation group to get to the bottom of that issue for the final report. They had to be summoned from wide and far after a lapse of seven years. The meeting also included top Intelligence officers, Dhafir, who was the head of Group 3 and responsible for Activity 3W that was under my direction, and Abdul Halim al-Hajjaj. It was an acrimonious gathering, as Halim feigned impartial silence. I was surprised, and alarmed, that he had to resort to such heavy weight Intelligence backing for his presentation, or lack thereof. A row developed between Dhafir and the two top Intelligence officers on sideline issues, such as the fate of the failed DIALOG station in Spain, which I was supposed to utilize. There were other claims and counterclaims, between Intelligence and Group 3, for who would be credited for acquiring some sensitive instrumentation. Exasperated, Jafar brought the meeting to a halt three hours later. We still had not gotten to the bottom of who had or what happened to the magneto optical discs. Halim managed to extricate himself from this issue like pulling a hair from a clump of dough.

In any event, when we did finish our final report and had submitted it to the IAEA, Jafar pulled out Khalid Rashid's request and signed his testimony that I truly harbored no further secrets on the nuclear weapons program as all secretes were exposed in the report that was just handed over to the IAEA inspectors. His assistant, the compassionate, efficient and professional Zaghloul Kassab made sure that the content of the clearance request was limited to very few eyes indeed. He even dispatched Omran, Jafar's trusted driver and my desert companion, to deliver the clearance with Jafar's signature agreement to Khalid Rashid. Omran refused to drop the mail at MIC, as is required, but insisted that he hand it personally to Khalid.

Khalid issued the clearance order to the Internal Security. He made sure to inform me beforehand of its date and number, so that I would in turn make sure that Abu Diyar's people would be present to receive that particular mail discreetly. Internal Security was nonetheless surprised that my name was to be released from their list of banned travelers, as its cross-referencing signaled that I might be worth more than what the clearance seemed to indicate, and replied to Khalid with a request for confirmation. Khalid Rashid immediately responded with a stern tone fully exercising his role and power over Internal Security. They relented.

An unexpected wall of bricks materialized. Abu Diyar could not finish the issuance of my passport because I was once issued a diplomatic passport. No civilian passport would be issued unless that diplomatic passport was returned. I had submitted it to the Foreign Ministry for it to be returned to the Passport Office when I retired. It had not arrived at the Passport Office.

I headed back to the Foreign Ministry and for many agonizing days dug into the administrative papers of two years ago. The Ministry had issued an official letter returning the diplomatic passport and attached the passport to it. Neither the letter nor the passport had arrived at their destination. It was yet another one of Salah al-Hadithi's poisonous agents in the administrative office who plucked that particular mail out in order to offer a final assurance to Salah that I would never be able to get a passport, having lost my diplomatic one. A copy of that letter, with a wad of cash, broke through that wall.

Abu Diyar visited us in late August with my authentic Iraqi passport and an exit visa awarded in it. His comment was all-telling. I would be indebted all my life to Khalid Rashid for this success. When I showed the passport to Khalid Rashid, he took it from my hand and kissed it. I bought him a car, which he gave as a gift to his brother. He boasted that he could do more for me, if needed, such as the release of Niran and the children's records. That attempt ended, at a later time, in a near fiasco.

The departure plan stipulated that we first had to go through the ordeal of getting Niran out to Amman, Jordan for her interview to secure the Canadian agreement for immigration. It would then be worth our effort to dash out, grab the immigration permit and immediately leave for

Canada, avoiding a prolonged stay in Jordan, which was rife with Iraqi Intelligence agents. One of the required papers that the Canadians had requested to be provided was a certificate from the police stipulating that Niran had not been indicted in any criminal offence.

We inadvertently committed an unwarranted folly.

We thought that this report was a simple matter and had innocently submitted, in the spring of 1997, the request for the certificate in a routine manner, to the great chagrin of Abu Diyar when he found out about that. We had just begun trusting him and telling him of our plans. We had not yet grasped the enormity of the net around us. He rushed to the police station to withdraw our request, but it was a day too late, as it had already been processed. A notice that we had requested such a certificate, which indicated that the applicant most likely intends to immigrate, was sent to Intelligence and to the Internal Security. The granting of an exit visa for Niran now, in addition to obtaining an exemption for her because of her Masters degree, also depended on the Security's explicit approval. In addition, her actual departure through the Iraqi border depended on the Intelligence Agency's approval.

In a period of one month, the Internal Security interrogated Niran and I three times. Each time, we would inform Abu Diyar beforehand of the intended meeting, which was commanded over a phone conversation, and he would rush ahead of us to appease the security officers, both with his influential contacts and a bit of cash. Sometimes we would dally around in front of the Security office until we saw his car leave the security premises before entering ourselves. To the security officials, our case became a milking cash cow.

It was much more difficult with the Intelligence Agency. I had appealed for help from al-Sahhaf. He obliged and directed me to an influential friend of his inside the Intelligence Agency, Mohammad al-Douri (Abu Omar), to track down and assist in processing the application. It was stuck with one of the most infamous, and a dire nemesis as far as my case is concerned, of Intelligence officers, who went by the pseudo name of Abu Muhannad. The cordial Abu Omar arranged for me to talk over the phone several times with Abu Muhannad. Characteristically, Abu Muhannad would repeatedly promise me that the approval would be sent the next day or the next week, but it would never materialize.

I had already gone through a long administrative procedure, with the assistance of powerful contacts at the University of Baghdad, to simply obtain the university educational records of Niran. This lasted for two weeks filled with continuous visitations and appeasements. A new law had forbidden the dissemination of even these records to stem immigration.

The next step was to get the approval of the Ministry of Research and Higher Education for Niran's exemption from the ban on traveling abroad. She had just passed the 45 years mark. I took her appeal to Humam Abdul Khaliq, the Minister of Higher Education and my former boss of twenty years at the Iraqi Atomic Energy. He ridiculed our modest attempt of an explanation at the time, namely, that Niran could make better money teaching in Amman. Nonetheless, and out of returning the favor for all the work I had done for him, Humam signed his permission as he expressed his belief that she would soon give up such futile effort and settle back in Baghdad.

Abu Diyar, relying on Humam's approval, and the Internal Security's clearance that he finally secured, had managed to obtain the one-month valid exit visa for Niran. It would, however, expire on the 7th of September. The first Canadian appointment for Niran in July 1997 was already missed, due to our folly of trying to get the police certificate the legal way. Subhi had delivered a letter to the Canadians in Amman pleading with them to await Niran's arrival before setting a new interview date.

The Intelligence stranglehold of Abu Muhannad was choking. His continued promises were intentionally dispiriting and dead-ended. After all the finagling, it would have been close to impossible to further extend the one-month exit visa for Niran. Abu Diyar, in his utterly reliable manner, came to Niran's timely rescue from the fangs of the Intelligence Agency. In early September, he asked again for Niran's passport, to which we had clung expectantly. Unbelieving, we gave it to him. He had managed, with a hefty sum of cash, to penetrate lower level Intelligence staff and forge an approval on Niran's file and had it sent to the passport office to be forwarded to the border point. Abu Muhannad, thinking that Niran's file is snuggly tucked away, was not aware of what had just happened right under his nose, until later.

Abu Diyar paid us a rare night visit and informed us that Niran should prepare to leave in a matter of two days. He promised her that I, with my readied passport, and the children would be following her a week after her departure. It was a time of heightened anticipation, as Niran's exit visa would have expired on the 7th of September. On the morning of Saturday, the 6th of September, a taxi stopped by our house very early in the morning, so as not to be noticed by the neighbors. Abu Diyar, in his car, was accompanied with his friend, Abu Hadil. Abu Hadil was going to Traibil, the exit border point to Jordan, to assume his assignment as head of the passport section there. Abu Hadil and Niran left in the taxi after a short portent goodbye. Once at Traibil, a passport officer objected to Niran's transit claiming that he had not yet received her Intelligence approval on his computer screen that would confirm Niran's written approval that she had brought with her. Abu Hadil interceded and ordered the officer to call the Baghdad office, where friends of Abu Diyar were waiting on the phone. They confirmed that Niran's approval was granted and that Traibil's data would be updated later that day, as it was. Abu Hadil then accompanied Niran all the way to the Jordanian border.

Elated, the children and I kept our mouth shut and prepared our suitcases. Abu Diyar was looking after the children passports' exit visas. Nofa was to get a new passport. She was originally added to Niran's when we got the passports in 1991 as Nofa was only five years old then. Abu Diyar, and his fellow security colleagues, had slightly modified Khalid Rashid's approval letter, appending the names of my three children to mine in order to facilitate the granting of their exit visas. They were ever so weary of other security officers lest they were planted among them by the Intelligence Agency.

Abu Diyar had set Thursday, the 11th of September, for our undeclared and expectant departure. I had arranged for the smoke screen excuses for my absence from my work at the UNDP and at our computer office with Husam. On Wednesday noon, as we were busy finishing packing our suitcases, Saad Yono, the husband of Niran's sister Nariman, walked in the door, his darkened face ashen. He took me aside and broke the devastating news to me.

Late at night, on the day before our intended departure, a personal messenger had arrived from the office of Saddam's personal assistant and trusted secretary, Abid al-Hamid Mahmood Himood, henceforth referred to as Abid Himood, with a dictum that was hand written and signed by Abid Himood himself, ordering the confiscation of the passports of Yamama, Tammam and Nofa Imad Khadduri.

The messenger grabbed the three passports and left. The stunned security officers immediately summoned Abu Diyar in the middle of the night. They hurriedly scurried around attempting to clean up any track of paper work that might lead to the intended doctored issuance of the exit visas for the passports of my children and, hence, their own complicity. There would most probably be a thorough investigation in the morning to find out if there was any wrongdoing at the passport office itself. They stayed up all night and did manage to clean up all the traces before the doors opened at eight in the morning. Abu Diyar was so shocked and bewildered that he visited Saad and asked him to break the news to me, as he did not even dare visit me. Abu Diyar also requested that I should stay away from him until he got to the bottom of what had happened and how the leak had occurred.

Throughly stunned, I took a good long look at my children and heaved in a very deep breath. I immediately realized the deep hole we had fallen into. That was when our painful and unwelcome odyssey started. All that had had transpired before then was a mere stroll in the park.

After waiting a few anxious days until Abu Diyar had satisfied himself that the breach was not from within his own circle, he confidently pointed either to a spying neighbor's report or to one of Niran's college colleagues who had tipped Abid Himood. What was most disturbing to him, though, was that my situation was so sensitive that it hair triggered this unheard of reaction. He again tried to convince me, while the escape routes were still manageable and controllable by him, to leave Iraq under false names and passports. Again, recoiling at the thought of being nameless refugees and probably risking the success of our immigration application to Canada with a changed name, I refused. I confidently promised Abu Diyar that I would try to recoup the children's passport through my own contacts and friends in high places. On the other hand, Abu Diyar was delighted that Abid Himood's emissary did not request my passport which indicated that Khalid Rashid's efforts had born fruit and that the Intelligence Agency was in fact not aware of it.

I broke the devastating news to Niran via a secure email.

She was staying with the generous family of Husam, my partner in the computer shop. I urged her to hold steadfast and find work until we could manage our escape. She did find a teaching post lecturing computer languages at a university. The Canadians had set January 1998 for her interview in lieu of the one missed a few months earlier. The children went back to school, Yamama to finish her baccalaureate (a nation-wide unified examination), the last and most difficult year in secondary high school and Nofa to her last year in primary school, yet another baccalaureate. I had managed, with a properly placed bribe, to switch Tammam to my previous high standard Baghdad College high school. I secured the services of a housemaid, the energetic Rita. She visited us twice a week to clean and cook a few days' ration at a time. Niran's extended family most honorably offered us immense assistance and support.

I began to slide into chronic depression.

Steadying my footing, I broke the news of the passport office raid to al-Sahhaf and boldly requested his assistance in recuperating the passports. He honorably obliged, in his steadfast manner. He again contacted Mohammad al-Douri (Abu Omar) at the Intelligence headquarters to intercede on my behalf. Abu Omar invited me to visit him at their headquarters. Its location was next to the Mansoor restaurant that was bombed by the Americans' second attempt to bomb Saddam before the occupation of Baghdad in April 2003, thinking that he was having a meeting there with his two sons, but killing 17 civilians instead. It is told that Saddam did enter that meeting place with his sons, leaving his personal guard at the front of the building. They then proceeded to leave the building immediately through the back door minutes before the exploding bunker busters. He apparently had purposely arranged to hold that meeting as a bait. He intended to confirm to himself the information that was reaching him indicating that the head of his guard entourage was in fact informing the Americans on his movements and location. It is rumored that Saddam summarily executed him after the failed attack.

At the insistence of al-Sahhaf, the puzzled Abu Omar had wished to get a straight story from me on what had happened and invited me over. I decided to take my passport with me to prove my case. Abu Diyar was most alarmed. Was I in my right mind to take my precious passport inside Intelligence Headquarters? It would be such a simple matter for them to strip me out of it. Nonetheless, I had to prove my point if there would be any hope of getting the children's passports back.

Abu Omar practically jumped from his chair when I declared that I had a valid passport. Unbelieving, he demanded to see it. I allowed him to merely look at it and jot down its number, clutching to it ever so tightly in a futile bravado attempt not to lose it if he chose to grab it. Upon hearing of Abid Himood's manner of intervention in the matter, he cooled my hopes to a minimum but still bravely promised to do his best, as a favor to al-Sahhaf. After figuring out the loophole through which Khalid Rashid had slipped me through, Abu Omar requested that I get another security clearance from Jafar, directed this time to Abu Omar personally, in order for him to authenticate and prove to others that my passport was really valid. He promised that the gesture would greatly improve my chances of recuperating the confiscated passports. The game felt like slipping the ladder rings and falling back down to the tail of the snake.

It took a lot of convincing to get Jafar to reissue his security clearance, now that my passport was public. He could have been implicated in assisting me getting it clandestinely. I had to reassure him that the passport was valid and official. Jafar wrote a most courageous and detailed defense of my security clearance, as far as the nuclear weapons program was concerned, even siding with my intention of leaving Iraq with my family, an indirect request to release my children's passports. I received it from him via Zaghloul Kassab who informed me of Jafar's resolve that this would be the last time that he would defend my immigration efforts. To add more resolution to that, Abu Omar had refused to receive Jafar's declaration from my hand but insisted that Jafar send it officially in a formal letter. Jafar was highly indignant by this request as his position, similar to that of a Minister, was thus challenged by Abu Omar. Nevertheless, he honorably did comply with Abu Omar's request, while casting a highly disapproving glare in my direction.

It was the infamous Abu Muhannad who was most angry, and pleased at the same time.

In all probability, he had found out that Niran's file, supposedly under his hand, was in fact compromised and resulted in a clearance that facilitated her crossing over to Jordan. The news of her departure probably came from a report from one of our Ba'athist neighbors, who lived five houses down the road, and was a known informer, or a colleague of Niran at the university. Another loud alarm was the news, whose source is still unknown to me, that her children were getting their passports. Basking in the false conviction that I was bereft of any passport, he thus succeeded in covering his failure of stopping Niran by getting hold of her children's passports. He probably contemplated that I might be trying to get a false passport and leaving through the Northern route, had I had the children's passports with their exit visas. Abu Muhannad most probably initiated Abid Himood's timely intervention. Abu Muhannad did not realize his immense luck in unknowingly stopping us from leaving the following day, not aware that my valid passport was ready and in eager hands. What was puzzling was Abu Muhannad's speedy access to Abid Himood, only a few days after Niran's departure, and his own relevance that induced Abid Himood to write the confiscation order of the passports by hand and to immediately dispatch his emissary.

I had spent thousands of dollars in trying to escape from Iraq. Half of that money went, through Abu Diyar, to all sorts of Intelligence and Security officers to keep track of my files as it moved around within their various organizations. To our knowledge at the time, there were eighteen different Security and Intelligence departments and agencies in Iraq, many of them keeping a watch on each other. Abu Diyar and I believed that I was being tracked very closely by three or four of them. Even then, Abu Diyar and I suspected that there was a super Intelligence Agency, higher than all of the eighteen agencies, that was directly controlled by Abid Himood. Furthermore, its members were ominously unknown to the rest of the Intelligence and Security services. Events, in my case, pointed to the suspicion of its existence several times. Abu Muhannad's lucky stroke in confiscating my children's passports was only one of them.

Al-Sahhaf firmly stood by my efforts and provided solace for my pain and depression, despite the apparent pitfalls that could reflect badly on him personally. He upheld my case even though I was no longer working for him but at the UNDP. He extended his influence to wherever his hand could reach. This included Abid Himood with whom he was on very friendly terms. (Abid Himood was captured on June 18, 2003). They would spend the late hours of some nights together, drinking and feasting. Sometimes I would be sitting next to al-Sahhaf in his office late at night, fiddling with his software, when he would get a call from Abid Himood over the phone for a pleasant chat. Al-Sahhaf greeted him reverently as "Doctor", since Abid Himood had received an honorary ineffective doctorate degree from the al-Bakir Military University that were bestowed at the time dime a dozen to people like Udai, Saddam's sadistic son, and Abid Himood. This close contact between al-Sahhaf and Abid

Himood stretched my hopes ever so thinly, to the supportive and infinite patience of Abu Diyar and his respect for my wishes and position.

The eminent Riyadh al-Qaisi, the Deputy Minister at the Foreign Ministry, also stood steadfastly by my side and extended his hand in firm support with my contacts at the Intelligence Agency.

With the solid support of al-Sahhaf and Riyadh's position on my case, Intelligence agents kept feeding me morsels of hope.

In the meantime, Abu Diyar had found a way to get close to Rafi' al-Dahham, the new Head of Intelligence (who was later poisoned to death, it is claimed). Abu Diyar had a very good friend who was in turn a very close friend of Rafi' al-Dahham. They were on such friendly terms that they would drop unannounced at each other's homes.

I had once met Rafi' al-Dahham, who was a former ambassador to Turkey, during one of his visits to al-Sahhaf's office. He had come to pay a cordial visit just before his promotion to be the Head of Intelligence. It was not uncustomary, in lieu of my close position with al-Sahhaf, that I would remain tweaking al-Sahhaf's computer while he entertained such visits.

Rafi' was so bombarded by Abu Diyar's friend, al-Sahhaf and Riyadh al-Qaisi on my behalf that he finally asked for my Intelligence file, convinced himself of my justified claims and took it by hand to Qusai, Saddam's son who was the nominal person in charge of Intelligence. Qusai reviewed the matter and announced that he saw no reason why the children's passports should remain withheld, but that since Abid Himood had ordered their confiscation, it was beyond his power to rescind that order. Such was the power of Abid Himood.

When al-Sahhaf and Riyadh visited Rafi' in hospital after his suffering a mild heart stroke, he informed them of what had happened with Qusai. I had already learned of that event through Abu Diyar's friend.

Niran's interview in January 1998 was approaching. Abu Diyar offered an alternative. He would issue me a passport with a different name; come with me to Amman to attend the interview with Niran and then return me to Baghdad. The whole trip would last a few days, after which the fake passport would be destroyed. He had enough friends to do that. Consulting with Niran, through encrypted email, about this course of action, she fell back on the promise of the Canadians that it would suffice for them meeting her alone. Hence, it was not worth the risk of this adventurous escapade. Her interview lasted ten minutes. They had asked for me. They claimed that we had misunderstood their intent in that letter and the application would not go through without my presence. They would not commit themselves to setting a new interview time.

The shock of that news, and the missed chance of securing the interview, delved me deeper and deeper into depression and despair.

Depression is when you cannot sleep. Feeling that you have nothing to say to anybody, you seek solitude in bed. The pillow becomes a source of torture, for as soon as you put your head on it, a multitude of thoughts, ideas, risk calculations, dire consequences and daily affairs splash around in your head like tumulus ocean waves that crash on every iota of brain cells that are kept awake because you are not tired or sleepy. This lasts for hours on end as you turn and toss on that accursed pillow. You refuse to get up in order to avoid talking to anybody.

Abu Diyar was my solace, tranquilizer as well as my savior.

Many an evening my children would notice me stealthily leaving home sullen and without even uttering a goodbye. They painfully felt my depression and left me alone to handle

Abu Diyar and my son Tammam in Sweden, where Abu Diyar
and his family managed to immigrate in 2003.

it. They knew that I was going to meet Abu Diyar, for I rarely left for anything else, trying to avoid people as best as I could. I would drive expectantly to the small two meters by three meters electric water cooler pump repair workshop of his brothers in the New Baghdad district. It was on a narrow side street, barely wide enough for one car to go through, with water running in a ditch through the middle of the street. The surrounding good souls in the bakery, clothes ironing and other small shops soon became familiar with my face and greeted me cordially. My spirit would either sink into further darkness or enlighten with lighthearted joy depending on whether I would see his old green Corolla parked in it usual place. If he were not there, I would be generously treated to sweet tea and wait for his arrival. Just seeing him walking toward me in his slow rotating manner, for he was plumb and rotund, was enough to lift my spirits and put a smile on my face that soothed my soul as we kissed. My children would sense my smile and start conversing with me again once I returned home.

Upon hearing the Canadian rejection of Niran's interview, Abu Diyar responded by offering a clever alternative in tune with my wishes. Perhaps I should take Khalid Rashid's offer of assistance in extracting new passports for the children, as he had had once promised? Khalid had weathered the fury of the Intelligence agents when they investigated how I had gotten hold of my passport. He showed them all of the proper documents that were issued according to their regulations. It was not his fault if they had such a loophole in their own instructions. I was not a General Director or higher. They let go of his throat but in typical Intelligence machination, installed an invisible human monitor, without his cognizance, to trap him in case he committed further transgressions.

All that was needed by Abu Diyar was another letter from Khalid Rashid's MIC Security office to the Internal Security mentioning that the lifting of travel restrictions from my name, as had already been decided, does cover and include the names of my children. A few months earlier, it was Abu Diyar's initiative to falsely append their names to my original release by Khalid Rashid that threatened to expose Abu Diyar and his colleagues when Abid Himood's envoy came to confiscate their passports. Now he wanted the authentic release to secure for them new passports.

I contacted Khalid and started to visit him again at his home across town in the rundown Shawaka district. During late 1997, the UN inspectors had treaded very closely to MIC and had applied enough diplomatic pressure to enter the premises. Khalid was ordered to clear MIC's security files and other records and relocate them. His home served as a transit point. At his leisure at home, he searched for and stumbled upon Salah al-Hadithi's poisonous letter with Saddam's terminating red ink dictum. With great risk, he tore it out of the file and shredded it. For that, I owed Khalid immensely as Abu Diyar again confirmed many months later.

Khalid was, however, disappointed with me for not securing a work position for his girl friend. I had to spend many days trying to get them to renew conversing with each other and dangling the work opportunity carrot once again in front of them.

Khalid prepared the required paper and called to inform me of its number and date. I immediately handed that information to Abu Diyar who headed to Internal Security to prepare his friends to receive and divert this letter accordingly.

That night Abu Diyar rushed to my home in a state of agitation, though he very rarely showed that demeanor. The number and date of the letter that I had given him was on the envelope of a most sensitive and highly secret document. Had my name been associated with that envelope in any way, I would have been forthrightly killed. How in the hell did this happen, he demanded?

I was in my car at Khalid Rashid's home at five o'clock the next morning, and waited nervously for him to get up. I confronted him at the door of his home as he left for work. He was shocked, and frightened. I drove him to MIC. He insisted that I wait for him in a far away car park, next to the Labor Union building. He walked back to MIC. In earlier times, I would confidently stride into the reception area of MIC and ask to see him. I waited for one hour in my car, chewing on my pipe and dragging its smoke like it was the nectar of life.

Khalid walked back to my car, his face ashen, withdrawn and sullen. "The bastards switched your letter in the envelope with the secret one after I had left the office last night at 6 o'clock", he explained. "I promise you upon my life that I will find out who has done this and I will kill him. From now on, I do not know you. Good bye". With that, I never saw Khalid Rashid again. The invisible monitor from Abid Himood had performed flawlessly.

Abid Himood was yet again one step ahead of Abu Diyar and me. Even had we managed, in February 1998, to get the new passports for the children, Abid Himood had initiated a diabolical trap. He had installed a new impenetrable barrier at the border crossing that would have foiled our attempt to leave. Abu Diyar explained it to me in depressing detail. Two air-conditioned kiosks, a mere two meters by two meters, were set up at the two entrances of Traibil, whether entering Traibil from Iraq or coming in from Jordan. Behind each kiosk was a waiting car with two lurking Intelligence officers watching the kiosk twenty-four hours a day. The engine of the totally tinted glass car was always running to keep the two officers either cool or hot, depending on the weather. Each kiosk had a personal computer with one operator. The computer contained a database of names that only and exclusively emanated from Abid Himood's office. The driver of each car that approached Traibil, whether leaving or entering Iraq, would need to dismount in front of the kiosk and hand over the passports of his passengers to the operator behind the computer who would do a name search at his own leisure. One glance from the operator's head to the driver of the car behind him would cause the idling car to screech forward, grab the inhabitants in the waiting car and drive them straight to Baghdad to Abid Himood's harsh hospitality. Ominously, the occupants of the kiosks and the idling cars were unsociable, and were not allowed to talk to the Internal Security passport officers or the other Intelligence officers at Traibil. It was truly a black hole of Abid Himood.

Abu Diyar felt that this new setup would greatly hamper any future attempt of ours to leave clandestinely. The trap around us was closing, and I had dallied around too long. I still clung to my hope in al-Sahhaf and Riyadh al-Qaisi to secure for me the confiscated passports of my children and that we would leave Iraq legally.

Abu Diyar had to smuggle a couple of Armenian doctors to Jordan. They were cousins of Hazmik. She is the delightfully extroverted wife of the magnanimous and noble Farouk Bazzoui, a distant cousin of mine. They were aware of some of the details of my attempts to leave as I had trusted them immensely. Hazmik requested the help of Abu Diyar for the sake of her cousins. Abu Diyar would only help Christians to escape, due to an unsavory religiously discriminating experience he had endured in the early nineties. Furthermore, he could only trust Christians to not expose him.

Upon his return from Jordan, Abu Diyar paid me a late night visit for a whiskey. This portended trouble. In Amman, he was approached with a proposal for me. He was asked to sound me on whether I was in accordance with the proposal or not. He delivered the message in a neutral and straight face waiting for me to decide. The Iraqi opposition group, belonging to Ahmad al-Chalabi, had approached him and asked him to inform me that they were willing, if I would agree, to pluck me and my children from Baghdad and smuggle us through the north. Somehow, they had figured out my plans and wanted to offer their assistance.

Ahmad al-Chalabi and I were schoolmates since kindergarten at Madam Adil. We were at the same third grade secondary class in Baghdad College in 1958 when his father spirited him and his cousins, Ghazi Alawi and Mahdi al-Bassam, to England to finish their studies. We had met again in Chicago during the sixties where he finished his PhD in Mathematics and in Beirut in 1965 to where I had often hitchhiked during these years.

Abu Diyar diplomatically put the offer on the table and waited for my response. I did not hesitate in responding. Since Ahmad's group was sponsored by the American Central Intelligence Agency, I would have no dealings whatsoever with them. Abu Diyar stood up, kissed me and told me that this was his conviction, too. Had I chosen otherwise and opted to consider the offer, then that would have been our last encounter. He assured me that, in the end, we would manage together. I slept without a pang that night.

There were few distractions from the gruesome depressive bouts that weighed heavily upon me in those days.

On Friday morning, I would take the children to the book market at Souk al-Sarai where they would wander around buying books and magazines that were laid out along the street. I would settle down on the street corner at Shabandar coffee house drinking tea, smoking a narghila (water pipe) and often met there and chatted with Denis Halliday, the head of UNOHCI, the UN organizational structure that administered the Oil for Food program. (The UNOHCI building suffered extensive damage, with many fatalities and injuries, from a large truck bomb on August 19, 2003). He was a down to earth, warm and expansive gentleman, easily approached and highly engaging. We had had several very frank discussions on the ailing electricity sector, my ambitions in leaving Iraq as well as on how things were going at UNDP. I offered abrasive words about its head at the time, Abdalla Auda. Denis did have similar impressions on this man but also wanted confirmations. At one café session, I brought Denis an illustrated book on the Marsh Arabs titled "Return to the Marshes" by Gavin Young. He was most pleased and promised to give me in return a pair of hand woven Kurdish sandals that I very highly value. He had acquired them from Twaila near the Iranian border and they were too big for his feet. I had worn their kind for ten years but had by then missed them because I could not get into these Kurdish areas anymore. Alas, I still await his gift.

My other relief and solace was my amiable, bountifully warm and very good friend Dhia al-Taie. His travel office, Blue Arrow, was a walking distance from UNDP. I would drop many times a week, in the middle of the morning. He would reach for the Backgammon board as soon as I entered his office. Without even a good morning salute, for he respected my sullen countenance, we would start playing. His assistant brought me tea and a narghila to smoke. After an exciting game or two, I would leave the same way I entered, but with a good bye. Then, he started dragging me to his closed poker clique every Monday night at the house of the dignified and humble aristocrat, Hikamt Tuaima. Dhia was most generous with his compassion. I inadvertently hurt his feelings later when he visited me in Amman when I hinted that I was so stressed out in my suspicion on who was leaking my information to the Intelligence Agency that I even dared, at one point, to think that it might have been him. After all, I was even suspicious of myself that I would talk about my secrets in my sleep and that would have somehow been recorded and transmitted to Intelligence. Dhia, however, took it as a grave personal insult and never talked with me again, until now. That remark is one of the very few things that I most regret in my life.

Then there was Farouk Bazzoui, my distant cousin, his wife Hazmik and his ever warm sister Amira, which means princess in Arabic, a name well befitting. We had an annual family tradition of traveling, in a caravan of several cars, to spend five beautiful days every Easter in the Kurdish north. We had memorable and unforgettable experiences with the company of our extended family. We visited every corner and village in the north over a period of many years. It was during such trips that I would make it a point to visit my nanny, Gozi, who was beyond ninety, in the Assyrian Christian village of Tilkaif. She had raised me for twelve years when I was

My nanny, Gozi, at her village of Tilkaif in the north of Iraq, 1993.

a child. I would repay her with a yearly visit with a small amount of cash to sustain her for another year in her small hut and surrounded by her neighbors, goats and chickens.

Many nights, I would take the children out to one of several places, especially after I would return from an uplifting meeting with Abu Diyar and feeling elated. We would either wander to Farouk Bazzoui's home, with Nofa executing her trademark knocks on their door, where we would always be most welcomed, best fed around their traditional round table filled with several small deliciously filled plates and best comforted. Or, I would take them to the Kurdish Sirwan restaurant where we would gorge ourselves on sumptuous *tikka* (lumps of lamb meat) and kebab (as well as a whole skew of grilled fat that we fought over) and freshly baked *khubuz* (unleavened wide circular bread). For a short dinner, I would take them to a nearby mobile restaurant, where fresh and exquisitely delicious *tikka* would be cooked in front of your eyes along with a drink of a lambs' yogurt.

In one of the laid-back gatherings around Farouk's deli-catessen tables, he turned to me questioningly. He was fol-lowing our stumbling efforts to leave. He was worried about my unwavering intent on escaping, with the children in tow, fearing for their safety. Would I really manage to get them out safely? Was it worth the risk of failure? Would I be able to find work to pay for their education? "If I, with my poten-tial, can not accomplish this, then who can?" was my response. "In sha'a Allah (God willing)" he replied, uncon-vinced. Yamama and Tammam, my eldest daughter and son, old enough to register what transpired, still recall overhear-ing that defying exchange.

Then there was Khoshaba's home comfort sanctuary. Rita and Mita (Mary) Khoshaba were Nofa's closest friends, as close as sisters. Their graceful Moroccan mother was a delight to be savored, along with her ever-optimistic husband.

Trying to frame a valid excuse for my legal exit from Iraq, I appealed to Jafar's younger brother Hamid, who heads an oil company in Sharjah in the Gulf, for assistance. During one of his visits to Baghdad, we met at Yahyia's home, Jafar's older brother, who was a staid moral support and is still a cherished friend. Hamid sympathized and obliged. A work contract to work for Hamid's oil company in Sharjah as a computer networking specialist was drawn up. Jafar's two sons, Sadiq and Amin, who traveled often to Baghdad, put in a lot of time, effort and risk to finalize the legal work contract, in the hope that it might convince the Intelligence Agency of my intentions. Amin even, very brave-ly, smuggled my scientific reports and articles to Niran in Amman during one of his trips.

Finally, my nemesis Abu Muhannad agreed to meet me face to face, after I had applied all sorts of contact pressure on him. His secretary arranged to meet me first at the Intelligence headquarters. I then followed his car to the rendezvous meeting place at the home of a hapless absent middle class owner, not far from the headquarters itself, who had lost his home to Intelligence for some imagined or real reason. I made sure of getting the license plate number of the secretary's car. Through that, I had hoped to have a lead to find out Abu Muhannad's real name. Had I his name, it would have allowed me to increase my leverage on him through powerful friends who would then recognize him instead of just Abu Muhannad, a pseudo name that was surprisingly unfamiliar to people in the know about Intelligence personnel. Abu Muhannad came 15 minutes later, and we settled down in an uncomfortable room with tea. His secretary left us alone. He was in his thirties, tall, powerfully built and very stern faced, though cordial, and explained during the initial ice breaking conversation that he had trained to be a pilot in Yugoslavia sometime earlier. Gazing at his cold dead eyes, one wondered how many people had suffered pain, if not death, under his hands.

It was one of the cruelest interviews that I had had to endure. He reduced me, uncontrollably, to tears as my pleas for the return of the passports of my children echoed dull from his thickened unfeeling conscience. I appealed through my fatherly responsibility towards my children; he responded with the government's prerogative to decide when to let us go. I pointed to the relics of antiquated scientists that we had been reduced to, fit only as museum pieces; he clung to the incorrect potential use, even to the last drops, of whatever vitality that could be squeezed out of us. In vain, I showed him my Sharjah work contract. It was a dead end, as far as he was concerned. He even faked surprise that Abid Himood might be able to do anything. That proved that he worked for him and that Abid Himood was my last chance. Abu Diyar was not able, with the car license number of his secretary, to track Abu Muhannad's name down. Abu Muhannad was more than an Intelligence officer. He remained an enigma to us, a part of Abid Himood's super security structure. It would not bode well, for him, if we ever meet again. I fell back on al-Sahhaf.

Niran was getting very lonely, desperate and even suggested that she should return to Baghdad. The Canadians had not corresponded with her since that failed interview in January and we did not know our status *vis-à-vis* them. I bolstered her resolve. I would not relent. I also sent her the few gold jewelry pieces that she valued in a tobacco metal container in the hands of Sa'ad, our trusted UNDP driver, who journeyed to Amman and smuggled it through the border. Sa'ad recounted Niran's bitter crying upon his return, to my deepening abashment.

The children were going through their exams. Niran's sister Nariman and her cousin Nada made sure that we got homemade food most days of the week. Even Niran's aged uncle and his warm wife would drop by in their thirty-years-old car driving all the way across Baghdad to bring us food every week or two. One of these times, it probably saved our lives.

Al-Sahhaf had left instructions with his secretary that I would be allowed to enter his office whenever I arrived at the Foreign Ministry, usually after nine at night, to tackle some of his ever sprouting questions on computer software or hardware. There, I would unburden to him, over a cup of Arabic coffee, my anxiety and depression and he would extend his continued attempts with Abid Himood for yet another day or two. It felt like dripping boiling oil torture. During that period, he had to travel abroad several times. The waiting bore unbearably heavy upon me. Niran was kept in daily touch with every movement of al-Sahhaf through UNDP's email. At UNDP itself, Abdalla Auda, the inept administrator heading the Agency, was aiming to kick me out any month. His Deputy, the compassionate gentleman, Peter Kouwenberg, held on to my leash as hard as possible. It was to Peter that I had, twice or three times, written on a piece of paper in front of him informing him that I may not show up in the office the next day, on my way out of Iraq. We did not dare even whisper of this in his office, lest it was bugged. Peter was the only person I would trust with this information in all of Iraq, save for Abu Diyar.

One night, in mid-July 1998, I paid al-Sahhaf my usual nightly visit. He had been promised by Abid Himood that my file would be cleared in the next few days, after repeatedly saying the same thing for the last three months. Expectantly, I entered and walked to his door. My ears perked up at an ever slight intonation in his secretary's voice asking me to please sit down and have some tea while al-Sahhaf finished a meeting with an ambassador. I controlled myself and sat down waiting for the tea. After a minute or two, I excused myself to go to the toilet. I headed straight down and out of the Ministry, never to return to it again. Abu Diyar was surprised to see me knocking on his door that night, as I rarely paid him a visit at his home to minimize any exposure of our relationship to his neighbors. I had left my car several streets further away.

My decision was made up. I was thoroughly convinced that al-Sahhaf was being lied to, unknowingly, by Abid Himood. Al-Sahhaf and I had been taken for a ride on a mirage promise. I relayed to Abu Diyar my heightened instinctive alarm by the tone of Sahhaf's secretary modestly requesting me to wait for five minutes. It was time that we should leave, clandestinely. I had given up completely on whatever clinging hope that al-Sahhaf would be able to get the children's passports back. Abu Diyar simply asked me to confirm my decision by next morning. He was, as always, ready to move.

A friend of my elder brother, Walid, called me on the phone. Abid Illah al-Tikriti, a deputy minister at the Oil Ministry, whom I had often visited to help him with his software, called from his office to tell me that he had just met my brother at a conference and that Walid was advising me against leaving clandestinely from Iraq, as it was a very dangerous endeavor, and that I might also not easily find any work abroad. My blood froze as I listened to Abid Illah recounting to me Walid's advice over his government phone. I immediately launched into a vehement denial and hurled insults against my brother for even suggesting these thoughts. Where in the hell did he ever get the notion that I might leave Iraq? Hefty curses fell upon my brother for even thinking of such imaginary routes for me. I vehemently confirmed that I do not need his advice, as I do not at all plan to do any such thing.

I knew that my outburst was not sufficient. What made Abid Illah commit such a flagrant security breach? Perhaps he had something to hide himself and tried to prove to whoever was listening that he was more of a patriot than I was. Abu Diyar was extremely alarmed. It took him several days, and a lot of extra cash, to verify that that conversation was hushed up and smothered.

A few days later, Abu Diyar instructed me to return my exit visa, which was simply a piece of paper that was attached to my passport for the past year, and collect its fee. It was a strange, yet very clever request. As I filed for hours at the financial government office responsible for returning the 400,000 Iraqi Dinars that we had paid as a fee for the exit visa, I am sure that I was being watched very closely by other interested parties, perhaps not believing their eyes. The whole administrative process lasted two days, but I finally contacted Abu Diyar informing him that I had annulled the exit visa and received my money back.

He instructed me to immediately deliver the passport to his brother's repair shop. That evening, he left a message asking me to bring 1,400,000 Iraqi Dinars in cash. Occupying several nylon bags, I handed him over the required amount.

He handed me back my passport with three adult exit visas and one child exit visa. He had added the pictures of Yamama, Tammam and Nofa to my passport and obtained one encompassing exit visa for all of us together. We would be leaving in two days, enough time for him to ready his friends in Traibil for our arrival. We still had Abid Himood's pitfall kiosk to cross.

On a Friday night, the last day of July 1998, I asked my children, yet again, to repack and close their suitcases, which had been open on the floor for eleven months, often emptied and refilled. They obliged respectfully, not daring to voice their doubt that they may soon be emptied again. I had arranged with Rita, our helper, that in the event of her coming over and not finding us, she should immediately move in with her family as guards for the house. I had even prepared, months earlier, the required papers stating that I am paying her a monthly wage for guarding my house.

On Saturday morning, the 1st of August, I took Yamama to enroll her at the University of Baghdad, being sure that our actions were being closely followed. I then dropped by the UNDP for an hour to inform them that I may go for a short visit to Mosul in the north and returned home. At noon, I decided that I should at least bid my friend Dhia al-Taie goodbye, for the persistent support that he had bestowed upon me, and measuring his sensitive demeanor. It was a very hot day, a few degrees above 50 centigrade. The first, then the second spare fuel pumps in my car failed, due to the excessive heat, within a few hundred meters from my home. I had to drag the car back into the garage and cancel Dhi'a's visit. I could not call him over the phone to say goodbye. I had to forego that gesture. Had fate allowed me to bid Dhia' farewell, he might have swallowed my inappropriate transgression upon his trust a few months later in Jordan. Alas, the damage was done.

Luckily, Niran's uncle just then called me over the phone to tell me that he and his wife would be driving to our home the next day in their old car to bring us some *kubba*, a wheat meal. I vehemently insisted that they should not bother to drive in that unbearable heat all that distance. I promised instead that I would be passing by their house by noon the next day to accept their *kubba*. He insisted upon his generous offer hoping to lessen at least one of my burdens. I finally convinced him of my promise to visit them with emphatic assurances. Apparently, that exchange itself had convinced whoever else was listening over the lines that we were at least going to spend the night in Baghdad, and hence, did not see the need to raise any alarms. I probably owe that phone conversation another breath of life.

I called Niran's sister, Nariman, and asked her to lend me their second car for the day. She obliged. I drove Yamama that afternoon to say goodbye to her close friend, Mina Ramzi, the only other person, aside from Husam our partner at the computer shop, who sensed our imminent intended departure as she hugged and kissed us.

Abu Diyar dropped by my home late in the evening bringing me a gift of a huge watermelon. I was outside talking with the neighbors. I immediately called for a knife and shared the watermelon. It was our signal that Abu Diyar had managed to secure all of the required arrangements for departure and that we would be leaving at three o'clock that morning.

Having gained the confidence of Farouq Bazzoui after smuggling the two Armenian doctors, Abu Diyar decided that we ought to leave from Farouk's home, and not mine, as ours was most certainly being closely watched all the time.

That night, I had, on separate round trips, surreptitiously transported the few suitcases to Farouk's home. Near midnight, I went to pick up Abu Diyar from his home. He would not allow me to be even seen by his own family nor allow me to bid his lovely wife goodbye. Instead, he met me at the end of his street. He had told them that he was going alone on a business trip to Mosul for a while.

There was one final matter to be dealt with. He had just been on the phone with Traibil. Abid Himood's shadow in the kiosk had been finally breached by Abu Diyar's friends, for the first time, and the official inside it had just announced that his final price is 1,000 dollars in cash. We stopped to buy one pack of Kent cigarettes, which was emptied and stuffed with 1,000 dollars, in 100-dollar bills, as agreed upon. It was a huge risk, as this was the first time that an arrangement had been sealed with this official. He could easily double cross us and get triple the amount as a reward from Abid Himood for turning us in. On the other hand, if he did go along with Abu Diyar's friends, he would gain their confidence and be assured of a steady stream of customers through that arrangement, and hence, secure measurably higher profits. They had to convince him, though, that I was neither a criminal nor from the government opposition. Furthermore, he had to be assured that my passport was official.

I dropped Abu Diyar at Farouk's home where the SUV with its trusted Jordanian driver were ready and loaded with our suitcases. I returned Nariman's car to her home, which was one block away from Farouk's. It was 2:30 am. The electricity was out and their neighbors were congregating in the street chatting and catching a cool breeze. I was very alarmed. Right in front of Nariman's house had moved Thamir Na'aman, head of the Physics Department under Jafar in PC3. It would have seemed very strange indeed that I would be returning Nariman's car at two thirty in the morning, and then walking gingerly away to the barking compliments of some stray dogs. Luckily, Thamir had slept through that night and did not partake in the neighborly gathering late that night. At 3 o'clock, we kissed

Farouk, Hazmik and Amira good-bye and sped off. Hazmik splashed buckets full of water behind us, for good omen. We had to show ourselves at Traibil's kiosk at 8 o'clock sharp when our waiting official would be assuming his morning shift.

Half an hour before Traibil, Abu Diyar, who was playing Backgammon with Nofa in the front, turned to me and seriously faced me. He said that we have half an hour together before crossing the abyss. He might as well confide to me two things that had he mentioned them to me in Baghdad, I would have definitely suffered a nervous breakdown and my will to leave would have faltered.

The first was my instinctive sensing of danger when al-Sahhaf's secretary asked me to sit down for tea for a few minutes. On the desk in front of the secretary was an order from Abid Himood ordering my arrest if I again entered the Foreign Ministry. Abid Himood was finally completely fed up with al-Sahhaf's continued nagging about my case. That order was just waiting for al-Sahhaf's perusal before coming into effect. I had left just before al-Sahhaf's obligation to put the order into action. By not returning again to the Foreign Ministry, I had done myself a great favor.

The second was Khalid Rashid's removal of the poisonous report of Salah al-Hadithi from my security file at MIC. Apparently, there was a cross reference to that report at the Passport Security. However, since they could not retrieve the original report, it was easier to delete the reference from my file. This helped immensely in processing my new exit visa including my children's. Had that report been preserved and brought forward with Saddam's red stricture on it, there would have been absolutely no way for Abu Diyar to accomplish what he had done with my passport.

At fifteen minutes to eight in the morning, we were at the petrol station watching the kiosk. At eight o'clock, the shift official walked in the kiosk and took his turn behind the computer. We drove up to the kiosk. The driver got down with the passports in his hands and submitted them to the office. A strange deadening calm had descended upon me. I felt numb but calm. It must have been my utter trust in Abu Diyar, who was keenly watching the official like a hawk. The official opened the passports, banged on his keyboard, smiled and returned the passports to the driver. We drove into the security zone of the passport complex. Abu Diyar closed all the curtains on the windows of the SUV and forbade us from descending from the car. The Jordanian driver was thoroughly surprised at this move, but did not offer any objection or comment.

Half an hour later, Abu Diyar came back with Abu Hadil, the same official who had gotten Niran out eleven months earlier and two other colleagues who were instrumental in securing Abid Himood's shadow at the kiosk. It was Abu Hadil's last month at this post, after which it would have been close to impossible for us to breach that exit point. Abu Diyar had deposited the packet of cigarettes with Abid Himood's shadow. The passport was stamped and ready. Through an open window, and without getting out from the car, as politeness dictated but overridden by Abu Diyar, they kissed me, one by one, wishing us well and bade us farewell. I will never in my life forget the warmth of their kisses. They probably could not believe that they had managed to get me through Abid Himood's net. We drove to the Jordanian side.

Jordan

The Jordanian Bedouin Intelligence officer called my name and herded the children, Abu Diyar and myself to an interview room. My passport was forged, he confidently claimed. Yamama was eighteen years old and Iraqi law does not allow a daughter of that age to be added on her parent's passport, but that it should be on a separate passport of her own. The attached pictures of the three children had no visible stamp to prove that I had not myself simply pasted them on the passport. The handwriting of the children's name was different from that of the original passport. We should return to Traibil.

I calmly denied any knowledge of any wrong issuance. Abu Diyar pointed out that we had just driven from Traibil twenty minutes ago. Surely, the Iraqi passport officers would have spotted any intended flagrancies and prevented us from leaving Iraq. Why not call them and find out? Abu Diyar had his own friends waiting by the phone connecting the two border crossings for three hours, just in case such troubles materialized.

The Jordanian officer disappeared for half an hour and returned having an "I found it" expression on his face. He confidently claimed that we were evading the Jordanian tax laws imposed upon those that stay longer than their legal residence permit in Jordan. That is why I had pasted the children's photos so as not to expose the fact that they have extended their stay in Jordan in order to avoid paying their due taxes.

Had this been Syria, I would have immediately interpreted his explanation as the need to dish out some cash stuffed in the passport as I flipped through it to double check what he was saying. However, being in Jordan, and he evidently being a Bedouin officer, I could not risk his wrath by suggesting such a bribe. I gathered that it was a face saving explanation for his detaining us for so long. Hence, I immediately rose to the occasion and produced Niran's legal permit card for working in Jordan, as a lecturer at a university. I modestly, and very humbly implied that he did not correctly interpret the situation as we were entering, not leaving, Jordan to visit the children's mother for a few weeks before returning to Iraq at the end of August for the start of their school year. He seemed relieved by the information.

A few kilometers inside Jordan, Abu Diyar turned to me and said, "Welcome to the land of Abu Abdallah", meaning the late King Hussain, the father of present King Abdallah.

A few days later, Abu Diyar and I were drinking and he confided to me: "I have never in my life prayed so hard to the Virgin Mary as when the driver went down to Abid Himood's officer at the kiosk with the passports in his hands". This revelation, coming from a very brave man, finally hit me as to the enormity of the precipice that we had just crossed. We had just managed to slip through the tiny eye of a very thin needle.

Within 24 hours from our departure, Abid Himood had sensed our departure and had dispatched five senior presidential palace staff members to Traibil. They arrived to Traibil on Monday at noon and had with them the recent photographs that I had taken of the children just two weeks earlier, in an off beat photo shop in Baghdad, that were then added to my passport. They fanned out throughout Traibil and went around asking every single passport officer at Traibil whether they have seen us recently, showing them the children's pictures and mine. Of course, they had not. We were in our SUV with withdrawn curtains over the windows. The few senior officers, who kissed me goodbye, swore that they had not laid their eyes upon me.

Before Abu Diyar's return to Baghdad, I handed him back my passport and asked that he would take it with him back to Iraq in order to make sure that all required stamps (some were water colored and could only be seen under certain ultraviolet light) were imprinted. I did not want to risk another potentially legal encounter when we would be leaving Jordan. He obliged. My passport returned to me ten days later with a taxi driver, bristling with all the required stamps.

Culminating the exhilarating reunion with Niran on Sunday August 2, an impromptu feast was thrown by the neighbors in the building where she had rented an apartment. I slipped away for a few minutes to fax a letter to the Canadians informing them that we have finally escaped.

Niran had not heard from them for seven months and we were not sure whether her immigration application was still active. They called us the next morning requesting our presence at the Canadian Embassy in Amman.

I went to the appointment first, as Niran had a chore to do with the children. She promised to follow me within half an hour. I was ushered in for a sole interview that lasted for three hours, while Niran and the children had to wait outside.

A senior Canadian diplomat faced me and announced, "I will ask you three questions. If you lie to any of them, you might as well forget your immigration application. Are you a Ba'ath party member? Did you become a General Director while in government service? Are you connected to Iraqi Intelligence or Security?" As I answered no to all three, the atmosphere relaxed and the Canadian generosity poured forth. I was most concerned with my safety while still in Jordan. So were the Canadians. He called me the next day informing me that he had obtained a special permission to fly my family to Canada within 24 hours, had I deemed it necessary to do so. The only problem would be that we would be offered a political asylum status. That was less advantageous and less favorable than a Landed Immigrant status, which would have to take its due course of six months. The decision was mine.

I consulted Abu Diyar. He phoned his friends back in Baghdad to get a sense of the Intelligence reaction, conversing in coded dialog. He learned of Abid Himood's hurried committee's investigation of our departure at the border crossing of Traibil. They were also intensely interviewing my colleagues at the UNDP. Peter, the Deputy, was advised by his weak hearted boss not to be in touch with me by email. Peter refused, upon which he was advised by a highly emotional boss that he would arrange, through the Ministry of Foreign Affairs, a *persona non-grata* status as a retribution, and thus force Peter to leave Iraq. Peter found that incredible, and rightly so. He faithfully remained in contact with me and visited me with his family on their way through Jordan several months later. Abdalla Auda attempted to block my last month's UNDP salary but for my threat of pressing serious charges against him for sexual harassment and smuggling. He finally relented upon the advice of fellow UNDP officials in Amman who took a cursory look at my claims.

My Jordanian partner at the computer shop, Husam, of Palestinian origin, who had no idea of how we had escaped, was also under great pressure, and sometimes physical. I later learned that it was the secretary of Abu Muhannad, who had driven ahead of me to that dreadful encounter with his boss a few months earlier that was tormenting Husam. He had immediately confiscated for himself whatever computers and printers in the shop that had belonged to me. He then sat himself in the shop with Husam for several days on end simply waiting for me to call and talk to Husam in order to glean any information on how I had left. Husam was also subjected to a harrowing interrogation ordeal. He was accused of supplying me and the kids with fake Jordanian passports. When his interrogation passed the psychological threshold and verged on physical harm, Husam had had enough. He simply packed up his suitcase one night and left Iraq for good, a week after our escape.

During that first week in Amman, I experienced several nightmares of volcanic suppressed fears. I would wake up drenched in sweat and trembling as I dreamt that I was somehow back in Iraq and being asked to report to Intelligence. I was beginning to unwind.

Abu Diyar also consulted a confidant who was a key Security official in the Iraqi Embassy in Amman. The Iraqi official could sniff Intelligence directives, but not Abid Himood's. Abu Diyar laid out for me the risks and the dangers of staying in Jordan. He was 50 percent sure of imminent danger and the possibility of serious harm befalling me. The decision was still mine. Niran and I decided to stick it out. We had sacrificed too much to end up as political refugees. I thanked the Canadians, but we would go through with the interview, medical examinations and the six months wait. In a week's time, I had found a teaching job at Amman al-Ahlia University, whose president was my friend and my brother's, from university years in the US. A telephone call to him was all that was needed. A resume in such situations was not that relevant. The name was sufficient to get a job immediately.

Dhafir Selbi, my life long friend and my previous boss at PC3 had by then managed, after seven years of waiting after his retirement, to obtain his passport. It was his first trip outside Iraq with the new passport. He gallantly paid me a visit, knowing fully well the wrath seething in Abid Himood's office at my slipping away. I had prepared a letter to al-Sahhaf before my departure. I had asked Farouk Bazzoui and his sister Amira to deliver it after my departure from their home on that fateful escape. They were reluctant. Undaunted, I asked Dhafir to take it back with him and hand it to al-Sahhaf. He obliged, as a true life long friend would, despite the risk of brushing the wrong way the frowning and vindictive Intelligence. It read (translated from the flowery Arabic):

Honorable Foreign Minister

Cordial greetings,

As I have left Iraq, I would like to extend to you my sincerest thanks and gratitude for the productive work environment that you have provided me while working in your esteemed ministry and for your sustained efforts in trying to lift the travel sanctions from the members of my family.

God knows and al-Sahhaf is a witness, as to how hard I have tried to cling to my country legally, with all my strength, patience, faith and devotion.

I shall stay outside my country for a limited time, but my love for Iraq will not die because I made sure that it is in the milk that was fed to my children.
Since the sixties and while I was a student in the United States of America, I took a decision upon myself not to marry but an Iraqi lady and that my children would be raised in the warmth and compassion of Iraq.

When I first applied for my retirement in the summer of 1991, and was asked for the reason by my superior, I made it clear to him that I knew that my request for retirement would take six to seven years to be processed. By then, I would have fulfilled my above decision and the time would have come for them to pursue their higher education outside Iraq, with all the strength that my wife and I would have left in us. We hope to return after accomplishing that, if we are still alive, God willing.

Despite the slow and bitter passage of the last valuable eight years, and in particular the agony of all members of my family throughout the last few months, during which I patiently attempted to fulfill all of the legal requirements to remove their travel restrictions, believing in the rule of the law, the efforts were, nevertheless, useless. Even though I have lost a lot of money, and a work contract in Sharjah, yet I have guaranteed the future of my children.

I will always remain that person who is ready to serve his country and willing to oblige its needs, in the field of my specialization. I wish to confirm to you that my tie to my country will not be severed no matter what distances and time might separate the two of us.

I would like to ask your honor one final request, hoping that it might be fulfilled.

I kindly ask you to request that a copy of this letter be inserted in each one of my security files, in the presidential palace, in Intelligence, in Internal Security, in the Military Industrialization Corporation and in the Iraqi Atomic Energy Commission files, for history's sake and for the future, in order to circumvent the attempts of any malevolent who might attempt to question my love for and my bonding to Iraq.

With peace,
Imad Khadduri
August 1998

Through the efforts of Niran's father, who had immigrated, along with his three sons, to the US years earlier, we received our immigration visa forms for the US while in Jordan. Niran was inclined to fork in that direction and rejoin her family. I had too much of a conviction and fear of the "American Way of Life" to allow my children to be condemned for life for what I felt was an alienated and violence driven society. I steadfastly refused to follow through.

Two months later, Abu Diyar returned to Jordan with more people in his toe. He broke the good news to me that Abid Himood had come to the conclusion, after intensive investigations, that a certain Kurdish driver by the name of Salah who worked for one of the UN agencies in Baghdad, had smuggled my children and me in a UN car to Arbil during one of his assignments. From there, I had slipped away through the north with faked Jordanian passports. Abid Himood had scribbled these facts on my file with the comment "Case Closed". I have no idea who Salah is, but whoever he is, I sincerely wish that nothing evil had befallen him as a result of this "investigation". Only then did I start to relax, for by then, we had yet again moved residence to delude tight surveillance.

The Canadian Embassy staff was very helpful. They had even telephoned the medical doctor who performed our medical examinations to courier our results for prompt approval. A senior Canadian diplomat finally traveled to Damascus, Syria to specifically bring back our urgently awaited for Landed Immigrant visas. Exactly six months after our entry to Jordan, we left for Toronto, Canada.

CHAPTER SEVEN:
Epilogue

Canada

Mississauga is not an easily spelled name, nor did we know anything about it when we put it as our intended destination in Canada. My uncle, the genteel and thoroughly generous Albert Abaji and his unforgettable wife Josephine Birjoni, the first lady to open a pharmacy in Iraq in 1946 and a most talented cook of Iraqi cuisine, lived there and we had put their address in our application form. My eighty odd years old uncle and his two sons, Namir and Ghadir, met us at the airport. Albert had insisted, in true Iraqi tradition, and by virtue of his being my uncle, that we have no choice but to stay with him at his apartment until we found suitable accommodation.

There was a severe snowstorm blowing outside the airport. Twelve years old Nofa started to cry profusely. "I don't have any friends here," she said. "I don't know how to speak English. I want to go back to Baghdad". She persevered and flourished.

We had chosen to immigrate to Canada for one main reason. We appreciated the fact that it would be our best chance to secure our children with their higher education relying on the quality of the Canadian universities and the support afforded by the Canadian government to university students, as well as free secondary school education. We had minimal financial resources of our own to spend on them for that purpose, otherwise.

We had provided a private English tutor, the venerable teacher Lawrence, for Yamama and Tammam throughout the last four years of our stay in Iraq. I had fully appreciated that

English would be the key to entry and sustenance at the university level. Foreign students had to get a good grade in the Test of English as a Foreign Language (TOEFL) for their admission to university. Even when Yamama had finished her secondary education in Baghdad, with excellent grades, we advised her not to enroll in a Jordanian university but rather spend the six months sojourn in Amman to attend an English language school. Furthermore, we had to rein her in when we arrived to Canada and made her repeat her last year at high school before entering university that she so eagerly anticipated. Tammam did finish his Canadian high school but failed to get sufficiently high grades in TOEFL to enable him university entry, though he could have entered a college. We decided that he should repeat his final semester in high school, and then take extra intensive English courses in preparation for the following year. He still did not get good enough TOEFL grades. Looking over his score sheet, I noticed that he had done well in the composition section, one of the five sections in the test. That was a sufficient indicator for me. Confidently, I asked him to repeat the exam a month after failing the first one. He passed with colors.

Finally settling down, I took a good look at the map to see where we were, having been too busy escaping and running before that. Not aware of the Canadian way of life, I was very apprehensive of coming to Canada at first, mostly fearing that I would find that its proximity to the United States of America would have resulted in it being a spit image of that "American Way of Life" culture, for which I had harbored so much antipathy.

Over the past four years, my reserve and guarded scruples had turned into a deep respect, admiration and joy at being an Iraqi-Canadian.

Being an avid newspaper reader for decades, I read the Toronto Star newspaper daily. The Star became my main venue for integration in Canadian society. Over time, the balanced reporting of events, the quality and professionalism of its columnists' views and the selected Letters to the Editor slowly gained my confidence, insight and admiration for the way Canadians think about themselves, their neighbors and their place in the world today. Their candor and self-criticism is a delight to savor. Their humbleness is their prime strength. Their enshrined "cultural mosaic", appended in 1982 to the Canadian Constitution as the Canadian Charter of Rights and Freedoms is living and prosperous and can be daily felt. Canada feels like an embodiment of a model United Nations country. Its quality of life is deservedly ranked among the top countries in the world.

I was once interviewed at the end of 2002, and before the invasion of Iraq, by an American radio show host. Her final question was, "You have studied, and lived for several years, at both US and British universities. How do you feel now knowing that both these countries are intending to wage a war against Iraq shortly?" My immediate response was, "I appreciate being in Canada." She was taken aback at this change of tact and quickly recovered: "How so?" she queried. I answered, "The American Way of Life is extracted from its Constitution which holds sacred three things, Life, Liberty and the Pursuit of Happiness. In my belief, that American pursuit of happiness is expended even at the expense of others. In Canada, their Constitution embodies three similar symbols as basic tenets to the Canadians. These are Peace, Order and Good Government". Silence ensued and then a thank you. She aired the whole half hour of the interview without that last exchange.

A few months after our arrival, two officials from the Canadian Security Intelligence Service visited us at our home. They were cordially invited in for coffee. I have been nurtured by previous experiences to keep my distance from such contacts as I have developed an allergy to Intelligence Agencies. They were well aware of my background.

Apparently, my presence and email was being monitored by an Iraqi who was regularly reporting on both to an official at the Iraqi Embassy at Ottawa. I was given the pertinent information and kindly requested to keep as low a profile as is possible. Confirming to them that their request was my already taken decision, I was emboldened to give them my career resume asking that they might perhaps assist me in finding a job at the Canadian Atomic Energy Commission. That did not bear fruit.

From a public phone, I called the number they gave me and asked for the person who was tracking me. After a few greetings and special Iraqi salutations, it was clear that he was an acquaintance of the Iraqi official at the Embassy. In no uncertain terms, and without revealing my own identity, I warned him of dire consequences if he would ever continue his surveillance and reporting, finally hanging up with ominous threats if he did persist, as I knew his address. I believe that probably ended his snooping spree.

I did keep a low profile concerning my past for three and a half years.

Coming out

Late in August 2002, I heard President Bush utter what appeared to be an ominous deliberate misinformation campaign. He claimed that Iraq was still harboring a rejuvenated nuclear weapons program and made it quite clear, as clear as can be mustered by him, that war may be waged on that premise, glancing over chemical and biological weapons. I turned to Niran and announced the end of my low profile slumber.

It took me two hours that same night to write my first article, *"Iraq's nuclear non-capability"*. The opening and ending paragraphs encapsulated much of what has happened over the next year, and perhaps for a longer time.

The first paragraph ominously pointed to "The war storm swirled by the American and British governments against Iraq, particularly the issue of Iraq's nuclear capability, raises serious doubts about the credibility of their Intelligence sources as well as their non-scientific and threadbare interpretation of that information. It is often stated that lack of inside information on this matter is scarce. Perhaps it is not too late to rectify this misinformation campaign".

The last paragraph predicted direly, "Bush and Blair are pulling their public by the nose, covering their hollow patriotic egging on with once again shoddy Intelligence. But the two parading emperors have no clothes".

The article first pointed out the glaring lack of any solid Intelligence information on Iraq's nuclear weapons program, which was in full swing for over a period of ten years in the eighties. This was evidenced by the failure of the sustained American bombing, during the 1991 war, of damaging but a few of the sprawling establishments dedicated to this program.

It then contrasted that lack of Intelligence information with the subsequent short sighted Intelligence on recognizing what had actually happened to the prime elements of any supposed rejuvenation of the nuclear weapons program; namely, its failure to portray the dismal condition of Iraq's nuclear scientific community, with many scientists and engineers unemployed and scrounging for work, during the nineties, the absence of their managerial leadership, and the consequences of the combined allied bombing and the UN inspection teams' demolition of our weapons program infrastructure, thus reducing any nuclear weapon hopes to rubble, except for archived reports and memories.

I sent the article for publication, in turn, to several prominent newspapers including the New York Times, The Guardian, The Washington Post, The Independent and The Times. As a reference for them to check on my credentials, I cited my University of Birmingham supervisors, Derek

Beynon and Malcolm Scott, who are still at their positions there. I received no reply over a period of two months despite countless email reminders. None had even bothered to contact Derek or Malcolm to confirm my authenticity. By the end of October, I sent it to the Toronto Star. At first, I had sent it to a prominent columnist there whose editorials I admired, but it was apparently missed in the volume of emails that he daily received. The next time I sent the article to the chief editor at the Toronto Star who forwarded it to Bill Schiller, the political editor. I received an immediate response with a request for an interview. The Star's investigative reporter, Kevin Donovan, rushed to meet me at the gardens of Seneca College campus, where I then lectured, and held a two hours interview. He promised me that the article, as well as a story on my background and credentials, would be soon published. I returned home to rudely find out that my home network security was seriously breached. I immediately called Kevin and requested a halt to any publication, until I restored my security. Whoever they were, they were very clever. They made me fully grasp the meaning of the word embedded. After many fruitless attempts to eradicate their sniffing and snooping, I had to format my hard disk and start my network afresh. It took me two months to restore a reliable (I hope) secure access to the Internet.

In the meantime, Sabbah Rumani, a Syrian friend of mine had suggested that I might consider publishing the article in yellowtimes.org, an independent news site that offers its "readers unconventional viewpoints from which to observe current events, and to encourage new thinking about the causes and effects of those events". Their platform appealed to me and I sent my simmering article to them. After a two weeks wait, I investigated the cause of the delay. It apparently had landed upon the plate of its chief editor, Erich Marquardt, but had again slipped through the cracks. A gentle prodding email produced the spark that ignited its publication on November

21, 2002[26] and the subsequent publication of eight other articles as events unfolded leading to and extending beyond the
occupation of Iraq; as well as attracting some woes for yellow-
times.org.

The volume of feedback emails to the first article was
astounding. Gratefully, none at that time was abusive or vile
probably due to the caliber of the yellowtimes.org audience.
A few days later, and emboldened by the volume of
response, Erich asked for another article. I responded that
the article that was just published covered my 30 years of
experience at the Iraqi Atomic Energy Commission and it
was not that easy to come up with another one. Erich was
tenacious. I then remembered that I had written a review of
the book written by Khidir Hamza in which he had fabri-
cated and exaggerated his importance in Iraq's nuclear
weapons program. I had published that article, written in
Arabic, two years earlier, under a pseudo name in Azzaman
online Iraqi newspaper published in London by Saad
Bazzaz. I had summarized that article into English later.
Erich volunteered to go over that short translation. He did
spend some time going over my summary and sent me his
version. Arabic is a powerful language. My article on
Hamza, in Arabic, was devastating, to say the least. Lest any
flavor is lost, I motioned Erich via email that I will under-
take an as exact a translation as possible of the whole article. He gave me a two hours deadline before updating his
site on the Internet. He had it by then and *"Saddam's
bomb maker is full of lies"* was published on November
27, 2002.

The article merely recalled snippets of experiences dur-

---

[26] A search for "Khadduri" in www.yellowtimes.org would return a list of
all nine articles, as well as the translation of some of them into other
languages. They may also be found, listed on one page, at these two sites:
http://www.reality-syndicate.com/syndicatepress/khadduri.htm and
http://www.redress.btinternet.co.uk/ikhadduri.htm

ing the two decades of working along with Hamza at the Nuclear Research Centre at Tuwaitha. His pompous, yet very cautious, postures kept him near the centre of the Iraqi atomic programs, yet shying from accepting any prominent and leading role. His sparkle to fame, quickly snuffed, came in 1987 when he initiated a report to Saddam Hussain claiming that the six years old nuclear weapons program under the direction of Jafar Dhia Jafar was faltering. This did lead to an overhaul in the method of work in the program and resulted in assigning to Hamza a leading role as the head of the atomic bomb design team. This responsibility lasted but for a few months as he committed a petty theft misdemeanor. He was cast aside and did not participate in the ramped extensive activities of the nuclear weapons program that engulfed about 7000 scientists, engineers and staff in the last three years before the 1991 war.

Hamza escaped from Iraq in 1994 and vainly sought refuge with the CIA and the two main Iraqi opposition groups based in London. His claim to being "Saddam's Bomb Maker" did not pass much muster with the American and British Intelligence agencies as the outlines of the gamut and history of the Iraqi nuclear weapons program came to light after the 1991 war, with no indication of his role in it. He ended up lecturing at a Libyan university where Fawzia, a lady CIA agent of Palestinian origin, monitored his activities. After his name was being used in several hoaxes and scam British newspaper articles[27], Saddam sent Hamza's son to try to convince him to return to Iraq. He rejected his son's appeal even though that must have been extremely dire to his family that he had left back in Iraq. He bolted to Rumania seeking yet again any refuge with the IAEA, the CIA and whoever would accept him. As fate would have it, his appeal coincided with Hussain Kamel's escape to Jordan

---

[27] *"Forged for heat of Iraq battle"*, by Solomon Hughes, The Tibune, June 19, 2003. http://www.tribune.atfreeweb.com/hughes13062003.htm

in the summer of 1995 where he had revealed details of some remaining hidden documentation of the weapons of mass destruction programs of the eighties. The CIA sucked him in, and without much brain washing, metamorphosed him into a willing and habitual prevaricator.

In his 1995 revelations to the UNSCOM/IAEA officials in Amman, Hussain Kamel[28] gave the following impression about Hamza's integrity:

"Prof. Zifferero—Dr. Khidir Abbas Hamza is related to this document.

General-Hussain Kamel—We call this person Hazem. He is dark, tall bigger than me. He is a professional liar. He worked with us but he was useless and was always looking for promotions. He consulted with me but could not deliver anything. Yes, his original name is Khidir, but we called him Hazem. He went to the Baghdad University and then left Iraq. He was even interrogated by a team before he left, and was allowed to go."

In January 1999, Hamza addressed the Seventh Carnegie International Non-Proliferation Conference[29]: "The plans were made and designed for an eventual production of 100 kilogram bomb—six bombs. That would be a reasonable arsenal in something like five to 10 years. So in a decade or so, Iraq would become a real nuclear power like Israel." In the light of the fact that Hamza was outside the nuclear weapons program during its final phase at the end of the eighties, and without our having even reached a "frozen" design for a bomb, this is at best a flagrant and fraudulent declaration.

---

[28] UNSCOM/IAEA Sensitive, Note for the File, 1995. http://casi.org.uk/info/unscom950822.pdf

[29] *"Who is Khidhir Hamza?"*, by Firas al-Atraqchi, in www.yellowtimes.org, November 27, 2003. http://www.yellowtimes.org/article.php?sid=888

Two years later, citing US Intelligence estimates, Khidir Hamza told nationally syndicated radio host Sean Hannity, "I don't think he [Saddam] has [nuclear weapons] right now but it may not take long for him to have it—a year or two probably. US Intelligence estimates at least a year. Germany estimates by 2005 are three nuclear weapons."[30] Hamza brazenly repeated the claim one week before Iraq's invasion in March 2003 on CNN's Crossfire TV show.

Assistant Defense Secretary Paul Wolfowitz is reported to have personally selected Hamza to join America's reconstruction team in Iraq. He is one of a group of Iraqi exiles who were to advise the troubled Office of Reconstruction and Humanitarian Assistance, now renamed the Coalition Provincial Authority headed by Paul Bremer since the summary dismissal of Jay Garner. Asked whether it was appropriate to send Hamza to Iraq after his association with forged documents, Pentagon spokesman Daniel Hetlage expressed confidence in his abilities. "Dr Hamza, who will be part of a team comprised of coalition partners, Americans and Iraqis, was selected for his extensive management experience in the nuclear field." Hamza has henceforth maintained a conspicuous and questionable public silence on the number of Iraqi nuclear weapons.

I was still weary of and involved in re-establishing my network security when an email arrived in December 2002 from a member of the Iraqi Action Team at the IAEA that was scouring Iraq for weapons of mass destruction in the fall of 2002. He had read both articles on Iraq's nuclear non-capability and on Hamza. He concurred with their content. We established an immediate rapport. He wondered whether I would be willing to have an interview with the IAEA and ultimately by UNSCOM, in accordance with the Security Council recommendations. I agreed on the condition that

---

[30] *"Saddam's Bomb Maker: Iraq Working on 'Hiroshima Size' Nuke"*, August 16, 2002. http://virus.lucifer.com/virus/2034.html

the interview would be carried out in Toronto, as I did not feel safe enough outside Canada. He tried hard to arrange that interview.

In the meantime, I was also persuaded by the suave approach of Michael Jansen, a journalist friend of my brother Walid in Cyprus, to partake with her some of my impressions and interpretations of the accelerating misinformation campaign being waged by the US and Britain. She published them throughout December 2002 and early January 2003 in various newspapers in Ireland, England, Jordan, Egypt and India. I was content to remain in my Canadian seclusion and security sending forth my opinions via her.

By the end of January, 2003 two events pushed me over the edge. I was angered by the cavalier manner with which the UNMOVIC team had invaded the privacy of Falih Hamza, a laser physicist who had no part in Iraq's nuclear weapons program. They claimed that the documents they had found in his home indicated that Iraq had a hidden laser enrichment process for uranium. Falih did carry out such research in the eighties. It did not bear any promise and he terminated the effort in 1988. We, the Iraqi nuclear team, even included that scientific experience in our final report to the IAEA in October 1997 in which we laid out the complete history of the nuclear weapons program. The documents and reports were neither secret nor related to the nuclear weapons program.

The other factor coalesced from input from both Michael and my friend at the IAEA. During January 2003, Michael was desperate to get into Iraq and was stuck in Amman, Jordan waiting for an entry visa. Informing me that al-Sahhaf would be passing through Amman on his way to a conference, I asked her to manage it so that he would see one of my emails. She did manage that through another journalist friend of hers who had an appointment with al-Sahhaf. Handing over my email, with indirect greetings to be extended to him, was sufficient for Michael to obtain her entry visa the next day. She was grateful and sent me a list of about three hundred sites that were visited by the UNMOVIC and IAEA teams. Those I sent to my friend in the Iraqi Action Team in Vienna for comment. He brushed the list aside and informed me of a revealing fact just one day before Blix's report to the Security Council on January 27, 2003. Upon Blix's insistence, the teams had obtained from the American and British Intelligence a list of about twenty five sites, one of which was ultra hush-hush. The inspectors duly visited and inspected each one of these sites in December 2002 and had found absolutely no evidence of any rejuvenated nuclear weapons program. In fact, some of them even came out stating that US Intelligence was providing them with nothing but "garbage after garbage after garbage".

Yet, Blix in his report to the Security Council on January 27, 2003 failed to mention the lack of findings in the ultra secret Intelligence information provided by the American and British Intelligence. He also promoted the case of Falih Hamza as being another belated uranium enrichment attempt by Iraq, hence adding fuel to the misinformation campaign. In all fairness, Mohamed el-Baradei, the head of the IAEA, did chide Blix the following day for not taking into account IAEA's knowledge on this matter, which was that the 3000 pages of documents were financial statements and Faleh's own lifetime research work, and had nothing to do with the nuclear weapons program.

Blix's omission and emphasis brought me out, completely.

I called Bill Schiller, the political editor of the Toronto Star and arranged for a meeting with him. The next day, Kevin Donovan rejuvenated and expanded his interview with me. A persistent, cordial and professional TV producer, Scott Ferguson from the Canadian City TV jockeyed with the Toronto Star for a first opportunity announcement of my coming out. On Friday morning January 31, 2003, the Toronto Star heralded a front-page article of my story[31] and City TV aired their Canada AM show at seven in the morning[32]. I was surprised to find out that many of my eight o'clock students at Seneca College had already either read the article or seen the interview.

A few days later, my IAEA friend from the Iraqi Action Team informed me in a private email that, "It would appear this is all a game and that no one is really serious about preventing a war. I have tried to be technically rigorous. I guess we will be swallowed up by the politicians. My plan to have a polite interview with you is a casualty of this mess".

We were inundated at home with TV crews and radio requests for interviews. Seneca College, most gratefully, protected my privacy there.

Some disturbing events portrayed the dark side of some American media. An Arabic named CNN representative, from their headquarters in Atlanta, had contacted me since the publication of my first article on yellowtimes.org requesting an interview. Having ducked my head while I recuperated from the security breach of my home network, she kept reminding me throughout two months that they were still interested in an interview, once I decided to come out. When I did finally do that, she conducted a pre-interview with me, to fathom what would I be talking about. She objected to my bringing up the neoconservative role in the build-up for of this war climate and other crucial, in my view, issues. I refused. She was vehement in her insistence.

---

[31] *"How Saddam plotted to get A-bomb power"*, by Kevin Donovan, The Toronto Star, January 31, 2003.
http://www.thestar.ca/NASApp/cs/ContentServer?pagename=thestar/Layou t/Article_PrintFriendly&c=Article&cid=1035777272571&call_pageid=96833 2188492

[32] CTV Television video clips, January 31, 2003
http://www.ctv.ca/servlet/ArticleNews/story/CTVNews/1044020356093_17///? hub=World

There was no further contact with CNN.

The American TV channel, CBS, had dispatched two staff from New York for a pre-interview. We spent a whole evening as well as the following afternoon together. They convinced their host of the "60 Minutes" show that I was authentic and had something to say that might be of value. They headed back for New York on the promise of filming the interview in two days time. The next day, they contacted me requesting that I may be further "evaluated" by an expert of theirs from Washington DC, claiming that he was an ex-UNSCOM inspector. We talked for about an hour. A few days later, I received the following email:

"The larger problem has to do with the weapons inspectors CBS hired as consultants. Steve Black, with whom you spoke, was impressed by your candor. He also believes you know a great deal about what happened up to 1991. However, Steve Black believes that once you indicated your desire to leave the country, you were no longer in a position to know for certain what the government of Iraq may have done with respect to maintaining and/or advancing the nuclear program. Steve Black believes there is evidence that the Iraqi government sought to preserve the nuclear program into the late nineties. He also thinks there is a great deal of classified information that the US Government is not willing to release. Given the obvious fact that his knowledge of the subject greatly exceeds my own, I am not in a position to refute his views.

I apologize for the time it has taken us to fully consider this matter. I have had several long conversations with Steve Black and his colleague, and it is clear that, in order to do the story, we would have to overcome the skepticism of the consultants. There is another possibility: that we will try to do a story about defectors such as Hamza and the information they have provided. But that is another story".

My painfully muzzled response was:

"Thank you for trying.

Pray tell, ask Steve Black, what was it that could have been (preserved) in the late nineties, aside from reports in the hands of the IAEA? Buildings, processing plants, phantom scientists, disappearing Calutrons?

Nice cover he warmth himself with, "a great deal of classified information which the US Government is not willing to release". Is it that worth "preserving" when they are being so seriously challenged by Blix and el-Baradei? Or is it the "intended" plan, as in the case of the "evidences" of massing Iraqi troops at the Saudi border and the "baby incubators"? Or perhaps his salary from the "US government" is high enough?

As to "once you indicated your desire to leave you were no longer in a position to know for certain what the government of Iraq may have done with respect to maintaining and/or advancing the nuclear program", how is it that I visited every single nuclear scientist and nuclear engineer while setting up my network across Iraq and roamed the Nuclear Research Institute till 1998? What a superhuman job they must have done to hide from me that evidence that is held in such a treasure chest such as the CIA?

This is a joke, no? A preserved pickle, no less.

My sympathies to the CBS.
Imad Khadduri

I had kept the American FOX News Channel at bay at the request of CBS in order for CBS to be the first American TV media to introduce me to the American public. When that flopped due to the above fiasco, a live interview on "From the Heartland with John Kasich" aired on Saturday, March 1 at 8

PM. The bias of the interviewer was, as I was warned of but still did not fully grasp, so flagrant that the interview quickly degenerated to shouting something like who the hell are you to defy the testimonies of Khidir Hamza and other Iraqi defectors who know that Iraq has nuclear weapons? My response to the effect that if they knew such evidences why don't they provide it to the IAEA and UNMOVIC that are on the ground in Iraq, was dimmed by his emotional outbursts at having somebody questioning his blind prejudice. His argument was not very dissimilar to that of CBS above. After Iraq's invasion by a few weeks and the utter lack of any signs of nuclear weapons, FOX had the gall to call again for yet another interview with John Kasich. I queried whether Kasich had another sleuth of defectors up his sleeve. In all fairness, Kasich did hold another interview and he courageously did correctly quote a few damning lines from my articles.

In order to prepare myself for the FOX interview, I had deliberately watched their channel for a few hours in my hotel room in New York on the night before John Kasich's show in order to grasp their slant. It was most disturbing to see an anchor on some FOX show who was trying to whip up the audiences support for the impending war by persuading the viewers with a most outrageous argument. He was expounding in great detail the amount of money, in billions of dollars, it would be costing the taxpayer if the US would decide to change its mind at that stage and withdraw the several hundred thousand American troops and their war armaments from around Iraq, hence, the argument projects, support the war. By virtue of being sent there in the first place appear to be cause enough for waging the war and in order not to loose money. This argument brought the horror of the "American Way of Life" vividly to my mind. It was a brazen effort to stem any opposition to the coming criminal invasion of Iraq. The anchor's position is in itself worthy, in my opinion, of a war crime tribunal.

In stark counterpoint, I would like to contrast the above instances with the openness, professionalism and generosity of Erich Marquardt of yellowtimes.org in Chicago. When his site was forced several times to shut down "for technical reasons" and for several days at a time after publishing some of my articles, Nureddin Mohamed al-Sabir from Redress Information and Analysis in England came to the rescue and offered to dedicate one complete page for a compilation of all my articles. Similarly, after publishing my article on *"The Mirage of Iraqi Weapons of Mass Destruction"*, Jason Cross from Colorado, US volunteered to host yet another well designed page for my articles, on his Reality-Syndicate site. (See footnote 26 on bottom of page 207)

To all of them I am very grateful and full of gratitude.

On February 5, 2003, the US Secretary of State Colin Powell performed a hollow and macabre, but laughable, show pretending to exhibit evidences that Iraq was continuing with its nuclear weapons program. *"The Nuclear Bomb Hoax"* was published in yellowtimes.org on February 7, 2003.

"The few flimsy so-called pieces of evidence that were presented by Powell regarding a supposed continued Iraqi nuclear weapons program serve only to weaken the American and British accusations and reveal their untenable attempt to cover with a fig leaf their thread bare arguments and misinformation campaign". Particularly irking was Powell's claim that Iraqi scientists were asked to sign confessional declarations, with a death penalty if not adhered to, promising not to reveal their secrets to the IAEA inspection teams.

Exactly the opposite was true. The four or five, as I recall, such declarations, the last of which was in 1997, held us to the penalty of death in the event that we did not hand in all of the sensitive documents and reports that may still be in our possession! One would have thought that had Powell's Intelligence services provided him with a copy of these declarations, and not depended on "defectors'" testimonies who are solely moti-vated by their self-promotion in the eyes of their "beholders", and availed himself to a good Arabic translation of what these declarations actually said, he would not, had he any sense been abiding by the truth, mentioned this as an "evidence".

This also applied to the "laser enrichment process" doc-uments that the UNMOVIC team theatrically uncovered in Faleh Hamza's home in January 2003, precisely because they were not part of the nuclear weapons program and hence, as far as Hamza was concerned, there was no need to hand them over to the government but kept them at home.

These same declarations also induced me later to throw a doubting conviction regarding Mahdi Shukur Ghali Obeidi's cache of the "centrifugal enrichment process" that he uncov-ered from under a rosebush in his garden in May 2003.

Expressing my doubt to the American physicist David Albright, an ex-inspector and President of the Institute for Science and International Security (ISIS) in Washington, D.C., who was also the conduit between Mahdi's coming out and the American authorities, I queried, "Why was Mahdi exempted from the penalties of our signing the four or five death declarations in case we hid something in our homes? Could it be that he in fact had Hussain Kamel's exemption to this order? In that case, why did Kamel trust nobody else when he undertook, over a period of two years, to hide many weapons of mass destruction documents him-self in his chicken farm, but not Mahdi's? Or did Mahdi take a huge risk and hid them to bargain his (and his family's) exit when Saddam is gone?"

David Albright, in a return private correspondence, con-firmed that the event was not staged and that Mahdi had Hussain Kamel's approval to hide them. The event still had not convinced Scott Ritter, the ex-inspector and ex-CIA agent who has fully turned around on the issue of Iraq's weapons of mass destruction. He claimed in an interview with Wolf Blitzer

(CNN) on July 11, 2002[33] "I believe you'll find that when you dig deeper into the Obeidi case, he's not telling the whole truth. Obeidi kept that material on his own volition."

Turning to Powell's claim that Iraq had tried to get nuclear grade fissile material since 1998 (he had side stepped the faked documents of the Niger "yellowcake" fiasco as it transpired five months later), a challenge to his "Intelligence" was posed. "Where is the scientific and engineering staff required for such an enormous effort when almost all of them have been living in abject poverty for the past decade, striving to simply feed their families on $20 a month, their knowledge and expertise rusted and atrophied under heavy psychological pressures and dreading their retirement pension salary of $2 a month? Where is the management that might lead such an enterprise? The previous management team of the nuclear weapons program in the eighties exists only in memories and reports. Its members have retired, secluded themselves, or turned to fending for their livelihood of their families".

Finally, "where are the buildings and infrastructure to support such a program? The entire nuclear weapons program of the eighties has been either bombed by the Americans during the war or uncovered by the IAEA inspectors. It is impossible to hide such buildings and structures. Powell should only take a look at North Korea's atomic weapon facilities, or perhaps even Israel's, to realize the impossibility of hiding such structures with the IAEA inspectors scouring everything in sight."

Powell needed only to have asked those on the ground, the IAEA inspectors delegated by the U.N. upon America's insistence, to receive "unavailable" answers to all of the questions above.

---

[33] The CNN interview with Wolf Blitzer on July 11, 2003 followed on the heels of Ritter's July 9, 2003 international press conference at Traprock. The entire conference is available for download and re-broadcast at http://traprockpeace.org/audio/ritter.mp3

Powell also wasted valuable show time on the aluminum tubes ruse promoted by the U.S. Army National Ground Intelligence Center[34]. A week before his speech, a report referred to the preliminary assessment of the IAEA that these aluminum tubes were not intended for nuclear weapons development but for the reverse engineering of 81-millimeter rockets[35].

Furthermore, new solid evidence has surfaced to counter the false claim of the aluminum tubes' use for a rejuvenated centrifuge process. In a hidden bit of news[36], also confirmed to me privately by David Albright, the previous head of the centrifuge enrichment process, Mahdi Shukur Ghali Obeidi, while asserting that he had hidden enrichment-centrifuge parts and documents on orders from Iraqi leaders for eventual use in rebuilding a bomb program, yet he confirmed "that since '91 they hadn't resurrected a nuclear weapons program", despite the pressures of American Intelligence. In the same cited article, Jacques Baute, chief U.N. nuclear inspector for Iraq, said that he had also learned of Obeidi's statements about the tubes and program status, adding that the Iraqi was in a position to know.

The White House had trumpeted the claim that Mahdi's hidden cache was evidence of Baghdad's nuclear bomb ambitions. However, US officials persistently failed to mention that Mahdi had in fact contradicted their assertions that the program had been revived and would need the aluminum tubes.

[34] "Speculation, fact hard to separate in story of Iraq's 'nuclear' tubes", By Bill Nichols and John Diamond USA TODAY, August 1, 2003. http://www.usatoday.com/usatonline/20030801/5374348s.htm

[35] "The Progress of UN Disarmament in Iraq: An Assessment Report", by David Cortright, Alistair Millar, George A. Lopez, and Linda Gerber, January 28, 2003. http://www.fourthfreedom.org/pdf/inspections_report.pdf

[36] "Bush staff put 'spin' in speech: Senator", The Toronto Star, July 18, 2003. http://www.thestar.ca/NASApp/cs/ContentServer?pagename=thestar/Layou t/Article_Type1&c=Article&cid=1058479813272&call_pageid=968332188854 &col=968350060724

My article ended on a painful note. Powell had claimed, "Let me now turn to nuclear weapons. We have no indication that Saddam Hussein has ever abandoned his nuclear weapons program." This verges on being humorous. But as the Arabic proverb goes: The worst kind of misfortune is that which causes you to laugh.

A week later, on February 14, 2003, the director general of the IAEA, Mohammed el-Baradei, presented his monthly report to the Security Council. Much to the chagrin, one would assume, of Bush and Colin Powell, the nuclear inspection chief's findings not only cleared the smoke from the imagined "smoking guns" of Colin Powell a few days earlier, but also dissipated some of the smog of misinformation with which the American government, aiming for war, had surrounded this issue. My effort to further elucidate his speech appeared in the article *"The Demise of the Nuclear Bomb Hoax"* which was published on February 16, 2003.

At the end of this article, attention was especially rejuvenated to Powell's unethical record on misinformation. He was certainly not new to this.

In *The Scourging of Iraq*, by Geoff Simons, "Washington lied persistently and comprehensively to gain the required international support [for the Gulf war of 1991]. For example, the US claimed to have satellite pictures showing a massive Iraqi military build-up on the Saudi/Iraqi border. When sample photographs were later obtained from Soyuz Karta by an enterprising journalist, no such evidence was discernible."

Simons refers to an article by Maggie O'Kane, published in the Guardian Weekend, 16 December 1995, which revealed that the enterprising journalist was Jean Heller of the St. Petersburg Times in Florida.

Eventually, the US commander—none other than Colin Powell himself—admitted that there had been no massing of Iraqi troops. But by then, the so-called evidence had served its purpose.

Yet with tongue in cheek, Powell claimed on February 14, 2003 in the Toronto Star, while still blistering under Blix's and el-Baradei's reports, that "force should always be a last resort—I have preached this for most of my professional life as a soldier and as a diplomat".

With the impeding onslaught, I had hoped that "Perhaps this time history should not be allowed to repeat itself."

A month later, on March 7, 2003, Mohammed el-Baradei , in his third report to the Security Council, thoroughly demolished and unequivocally disproved most of Colin Powell's alleged 'evidences' of Iraq's continued nuclear weapons program after the end of the 1991 war— 'evidences' which Powell so brazenly offered in a theatrical presentation to the same Security Council just a month earlier. It was time to name things by what they are. *"The Fig Leaf of Moral Impotence"* was published in yellow-times.org on March 10, 2003.

It had taken the IAEA experts only a few hours to determine that the Niger documents were fake, a feat not beyond the abilities of the Intelligence sleuths. Yet it had to go through the misinformation grinder of the neoconservative Intelligence cooks and put on a plate to Bush to pronounce in his State of the Union address.

Unabashedly, Bush had given a speech on the same day, March 07, 2003, portraying the gathering dark clouds of a criminal war against Iraq, in the terms of a poker game. He challenged other countries opposed to the criminal war to "show their cards" while the US and the U.K. would conveniently keep their cards hidden.

In the event, another warning in the article was missed. "Lest he misses the point, he is playing a game of Russian roulette, and his fig leaf has fallen," I predicted.

The day following el-Baradei's speech, his assistant, Jacques Baute, who is the Head of the Iraqi Action Team of IAEA inspectors, did finally fly over to Toronto for a cordial meeting. We knew each other from Baghdad, as he was an inspector in Iraq for many years. We spent seven hours reminiscing mostly about the nuclear weapons program before 1991, as he was well versed on its non-rejuvenation after that time. I still had with me a few bits of information on that period that fitted in his jigsaw puzzle. He encouraged me to persuade other colleagues to come forward to bolster the IAEA's position that it had interviewed senior scientists in the program. I partially succeeded as the bitterness to the callous and arrogant treatment displayed by the inspectors, in addition to the exuberant fees paid to them, was deeply felt; especially by those who are still in Iraq.

The loud drums of war silenced any truth. The crescendo of lies heightened. On March 16, 2003, Vice President Cheney, one day before the 48 hours deadline announced by Bush for his invasion of Iraq, had the audacity to blatantly claim that IAEA's Mohammed el-Baradei is wrong and that US Intelligence had proof that Iraq harbored a rejuvenated nuclear weapons program. To pre-empt any display of their customary bag of tricks, *"Cheney's Bogus Nuclear Weapon"* was published on March 19, 2003 ominously pointing out that the invading forces will not find any trace of a rejuvenated nuclear weapons program, save for any bogus evidence that the Americans themselves might place once inside Iraq.

Weeping, at the moment, for Iraq

An unfair reaction developed to the above articles by close Iraqi and Canadian friends. Their attitude prompted me to write the following *"Postscript"* to the published articles. It was sent only to selected Iraqi and Canadian friends on March 13, 2003, one week before the invasion of Iraq:

"There is the assumption that if one decries the coming slaughter of Iraqis by Americans, that one must also curse Saddam at the same time; otherwise, so the accusations go,

you are defending Saddam. This self-imposed condition may be summarized in the following anguished retort of an Iraqi friend: "I watched the video of the whole interview. Very interesting, but what I also find interesting is the fact that you never said a word about Saddam and the horrors he caused the Iraqi people!! Not a single word!!. . I think any Iraqi should never miss an opportunity to tell the world what Saddam has done to Iraq and the Iraqi people."

Both my Iraqi, and non-Iraqi, friends have made similar comments as above, and have attempted to coax me into denouncing Saddam each time I discussed Iraq's nuclear non-capability. However, this suggestion by my friends merely reflects the dead end of their own logic, as will become evident in the following exchange[37].

I personally can touch and feel the pain of my Iraqi friends. They have suffered much more than my family and I at the hands of Saddam's Intelligence and Security apparatuses. We had left Iraq without their knowledge or their approval and at the risk of our death. Many others have lost their relatives or families. I accept their narrow vision of the need to curse that reign of terror daily, to spend nearly every breath venting their deep anguish and anger; however, if they could cast their justified emotions aside for just one moment, I ask for them to realize that the Americans are utterly lacking a viable plan for Iraq and the Iraqi people after they drop their hundreds of bombs and fire their destructive missiles at Iraq. This will lead to at least tens of thousands of dead Iraqis and the easy defeat of the tattered remains of the Iraqi army. **Iraq and the Iraqi people will be in a state of free fall, dropping into a deeper abyss,** *with Turkey, Iran and Israel (with its own agenda against the Palestinians) all*

---

[37] *"A Warmonger Explains War To A Peacenik"* (Anonymous author), posted as Postscript to the five articles... in http://www.reality-syndicate.com/syndicatepress/khadduri.htm

*eyeing pieces of Iraqi flesh to bite off. The oil has already
been marked.*

*I do not want to counter right now this same stance held by
the non-Iraqis, whose selfish shortsightedness reveals deeper
antagonisms.*

*In both cases, I reject their bankrupt final solution to their
one-dimensional way of thinking, that "if you do not agree
with us, why did you leave Iraq and why not return to Iraq".*

*The coming war was not launched in the spur of the
moment. It is an opportunity seized upon after September of
1991. Its seeds have been planted since the early nineties by
a clique of American neoconservative right wing thinkers,
with strong sympathies (and some, even ties) to Israeli interests. These thinkers engineered their plans for reshaping the
Middle East through their work with the American
Enterprise Institute, the Project for the New American
Century, and other like-minded organizations.*

*They have managed to ascend to high ranks in the
American State Department, the White House and especially
the Pentagon. Among these neoconservative thinkers are
Richard Perle (Chairman of the Pentagon's Defense Policy
Board), Paul Wolfowitz (Deputy Defense Secretary),
William Kristol (Chairman of the Project for the New
American Century), Douglas Feith (Under Secretary of
Defense and Policy Advisor at the Pentagon), Lewis Libby
(Vice President Dick Cheney's Chief of Staff), and others.*

*A search on the Internet will reveal much more of their
thinking, agenda and reports. Visit some of these organizations' websites, at http://www.aei.org/ and
http://www.newamericancentury.org.
Despite fierce American media support, their arguments are
being riddled with holes, yet the war crimes they plan on
committing in Iraq will still take place.*

My five articles, and the numerous TV and radio interviews, are solely intended to shred even further their flimsy arguments and expose the extent of the misinformation that is beamed to the American people and others to blind their vision of what is actually being enacted.

These neoconservatives will, sooner or later, be fully exposed and cast aside; hopefully they will appear in front of an international war tribunal along with Saddam.

The Iraqi people will resurrect".

Within about three weeks after the occupation of Baghdad, **"The mirage of Iraq's weapons of mass destruction"** was published on April 30, 2003.

Having begun to realize that Iraq had indeed destroyed its entire stockpile of chemical and biological inventories and warheads in 1991, as affirmed in the suppressed testimony of Hussain Kamel to the IAEA inspectors in Amman, Jordan eight years earlier, my conviction that there will be no discovery of any weapons of mass destruction was cemented by the confirmation, during the short TV interview of Amer al-Saadi, the most senior scientific advisor to the Iraqi government, who gave himself up to the American occupation forces a few days after securing Baghdad with the words "Iraq does not have chemical and biological weapons of mass destruction. I have nothing to hide. Time will bear me out."

Indeed, time is bearing him out to the chagrin of Bush and Blair. The American and British hopes of finding any weapons of mass destruction in Iraq, not planted by them, are vanishing mirages.

The article concluded, "Bush, Blair and their senior officials lied to their people, knowingly, and waged a criminal invasion in lieu of this reason. Is this the democracy model for a "liberated" Iraq?"

In an article to be posted on a newly launched English al-

Jazeera Arabic TV channel Internet site, *"No Iraqi WMD? What were the 13 years of sanctions for?"* an attempt is made to put a face on the disfigured façade of Intelligence misinformation that has preceded the occupation of Iraq.

The western (and Israeli[38]) Intelligence communities, having missed the extent of the extensive nuclear, biological and chemical weapons programs during the eighties, have missed yet again the destruction of these materials after 1991. Their analysts were not prepared to consider Iraq to be capable of taking the step of destroying its weapons of mass destruction in 1991. Tellingly, and in both instances, they lacked Intelligence collection from human sources in Iraq and first hand accounts of what was being undertaken before 1991, and what has happened after the 1991 war.

They are still investigating weapons of mass destruction mirages conjured up in Intelligence task forces and groups.

The Americans have been searching Iraq thoroughly for weapons of mass destruction for the past many months, even before their invasion of last March. Barton Gellman reported in the Washington Post on June 13, 2003 that "A covert Army Special Forces unit, operating in Iraq since before the war began in March, has played a dominant but ultimately unsuccessful role in the Bush administration's stymied hunt for weapons of mass destruction, according to military and Intelligence sources in Baghdad and Washington. [This was] Task Force 20, whose existence and mission are classified and is drawn from the elite Army special mission units known popularly as Delta Force" Its principle mission was to "seize, destroy, render safe, capture, or recover weapons of mass destruction". Earlier reports indicated their presence in Iraq since early February.

---

[38] *"Trojans, journalists, spies and puppets"* by Paul J. Balles, 16 July 2003 http://www.redress.btinternet.co.uk/pjballes10.htm. The article refers extensively on two books by Victor Ostrovsky *"By Way of Deception"* and *"The Other Side of Deception"*.

They were not able to find a single shred of active evidence of Iraqi weapons of mass destruction despite their impressive support facilities and detection capabilities, including roving biological and chemical laboratories. They are still searching, in their futile attempt, for more frustration.

Task Force 20 was followed in early April 2003 by more than 900 specialists in the publicized 75th Exploitation Task Force who also "found no working non-conventional munitions, long-range missiles or missile parts, bulk stores of chemical or biological warfare agents or enrichment technology for the core of a nuclear weapon" that were specifically cited by the Americans as part of Iraq's concealed weapons of mass destruction arsenal. They left Iraq in early June empty handed and thoroughly dismayed.

After September 11, 2001, Paul Wolfowitz, the Deputy Secretary of Defense, had established a small super Intelligence group, who self-mockingly called themselves the Cabal. It was mainly staffed by ideological amateurs to compete with the CIA and its military counterpart, the Defense Intelligence Agency.

The agency, called the Office of Special Plans (OSP)[39], was set up by the Defense Secretary, Donald Rumsfeld, to second-guess CIA information and operated under the patronage of hard-line conservatives in the top rungs of the administration, the Pentagon and at the White House, including Vice-President Dick Cheney. They relied on data gathered by other Intelligence agencies and on information provided by the Iraqi National Congress, INC, the exile group headed by Ahmad al-Chalabi.

The director of the Special Plans operation is Abram Shulsky, a scholarly expert in the works of the political philosopher Leo Strauss, the guru of neoconservative

---

[39] *"The spies who pushed for war"*, by Julian Borger, The Guardian, Thursday July 17, 2003
http://www.guardian.co.uk/Iraq/Story/0,2763,999737,00.html

thought. He served in the Pentagon under Assistant Secretary of Defense Richard Perle during the Reagan Administration, after which he joined the Rand Corporation. By last fall, the weight of the Office of Special Plans had rivaled both the CIA and the Pentagon's own Defense Intelligence Agency as President Bush's main source of Intelligence, as it funneled and cherry picked information, unchallenged, regarding Iraq's possible possession of weapons of mass destruction and the alleged connection with al-Qaeda.

A great deal of the bad information produced by Shulsky's office, which found its way into speeches by Rumsfeld, Cheney and Bush, came from al-Chalabi's INC. The INC itself was sustained by its neoconservative allies in Washington, including the shadow "Central Command" at the American Enterprise Institute.

How is this deceit fiasco handled by the neoconservatives? Pass the buck.

In mid June 2003, the White House recanted and put CIA's Tenet in charge of the search for the faltering weapons hunt shifting the responsibility away from the Pentagon.

The CIA had already downgraded its own two specialized teams on Iraq, the Iraq Task Force, a special unit set up to provide 24-hour support to military commanders during the war, and the Iraq Issue Group which is responsible for the core analysis of all the Intelligence the United States collects on Iraq. A senior official of the former was reassigned to the CIA's personnel department and the head of the latter was dispatched on an extended mission to Iraq.

According to Greg Miller of the LA Times on June 14, 2003 "two of the key players on this problem have essentially been sent into deep exile," said an agency official, who spoke on condition of anonymity. The official added that the changes seemed designed to show the administration that "we're being responsive to charges that we did not perform well".

In taking over the search from the Pentagon, CIA Director George Tenet will have direct responsibility over a newly created Iraq Survey Group, which is now in Iraq, to "significantly expand" the so far fruitless hunt for chemical and biological weapons. As a last straw, Tenet adroitly passed along the baton to David Kay, a former UNSCOM inspector in 1991, to serve as a "special advisor" to the newly formed 1400 strong team, and to be in charge of "refining" the overall approach for the search for Iraq's weapons. Kay's locked mindset that muddled his professionalism and objectivity regarding Iraqi weapons of mass destruction may be surmised in the following position, "For me, the real change occurred in '94. By '94 I was no longer an inspector, but I was testifying and writing on Iraq that 'There is no ultimate success that involves UNSCOM. It's got to be a change of regime. It's got to be a change of Saddam.'"

By that time, all that was left of Iraq's weapons of mass destruction programs were reports, memories and ruined establishments. Hussain Kamel, who headed all Iraqi weapons of mass destruction programs, had attested to that in his testimony to Ralph Ekeus, who was in charge of UNSCOM in 1995, before Kamel's death in 1996. Kamel's testimony was suppressed for eight years until it's ferreting out in February 2003.

David Kay, who went to Baghdad in July 2003, may soon be jarringly awakened to the reality of his sustained misinformation on Iraqi weapons of mass destruction with the coincident capture of Abid al-Hamid Mahmood Himood, Saddam's most trusted secretary. Abid Himood was in fact in total charge of preventing the UN inspection teams from encroaching upon Saddam's palaces and private space. He was privy to relevant communications with Iraqi officials during their encounters with UN inspection teams that scoured Iraq before 1998 and the rejuvenated inspections in the fall of 2002.

Abid Himood will be only collaborating the claims of Amer al-Saadi, the chief scientific consultant to the Iraqi government who surrendered to the occupation forces in mid April 2003. Upon his incarceration, al-Saadi maintained that Iraq had had no weapons of mass destruction for the past ten years. Himood will confirm that this was in fact true.

Aside from Abid Himood, Qusai Saddam Hussain was the most knowledgeable government official who could have confirmed the existence, or the more accurate lack thereof, of any Iraqi weapons of mass destruction as he was in nominal charge of all military and civilian Intelligence and Security organizations.

The seemingly blind rush on killing Qusai in Mosul on July 21, 2003 instead of capturing him alive as in any hostage situation (without even any hostages) casts further shadow on the questionable intent of the rushed American bravado.

David Kay has probably interrogated Himood and Amir al-Saadi by now. Kay cannot anymore claim to be stymied by the tricks and subterfuges of an intact Iraqi regime. An excruciating hyperbola of reasoning is now expected from Kay, who is not a scientist, as he fails to solve the maddening mystery of the elusive Iraqi weapons of mass destruction; and the circle has come around to close. The truth would be in order, for a change.

"Iraq's free fall" was published six weeks after Baghdad's occupation, on May 23, 2003. The pain and the enormity of the damage were colossal.

America's threadbare credibility on Iraq's weapons of mass destruction is stretched to the limit, and is snapping in the face of those that purveyed faulty Intelligence filtering, intentional manipulation and double-speak excuses. It highlights the immensity of the crime perpetrated against Iraq.

Tragically, the occupation of Iraq has indeed thrust the Iraqi people into a state of free fall. The despotic lid of Saddam was uncovered but the hole is deep and unpredictable.

The centers of power in central and southern Iraq, the ministries, banks, courts, hospitals, schools, import/export structures, and decision centers are mostly demolished. Daily living conditions, on top of a very serious lack of law and order, are the lack of electricity, water, medicine, petrol, and salaries. With no salaries being paid to millions of civil servants for over a month now, thousands of families have been reduced to destitute levels in masse, fuelling further lawlessness.

The Americans have only proved that they excel in mobilizing armed forces and in dropping missiles and bombs, but not how to prevent a society from disintegrating. American "Intelligence", by the time Bush speaks it, is so flawed that it is pathetic. In addition to their delusions about weapons of mass destruction, Bush and his neoconservatives are nothing but war robots and see the world only through US goggles. Preventing a society from disintegrating is not on the agenda or the intent of these neoconservatives. On the contrary, it is in fact deliberately envisioned in order to be rebuilt according to their *Pax Americana* vision[40]. The Americans and the British have, in a sense, raped Iraq and left it bleeding. There they stand, with their pants down, licking their lips, just waiting to manipulate the oil industry.

The article concluded, "A month and a half after the end of hostilities, and in the face of unexpected and mounting Iraqi resistance, the invaders have finally submitted and managed to pass a U.N. resolution proclaiming that they are not the liberators they claimed to be but are, in fact, the occupiers of Iraq. They should only expect what occupiers deserve."

Belatedly, and aside from widespread use of cluster bombs on civilian populated areas[41], including the cities of Baghdad, Nasiriya, Basra, Hilla and Rashdiya[42], we learn of

[40] *"A Bush vision of Pax Americana"* by Gail Russell Chaddock, The Christian Science Monitor, September 23, 2003.
http://www.csmonitor.com/2002/0923/p01s03-uspo.html

[41] *"Use of cluster bombs in Iraq criticized"*, by Frances Williams in Geneva, May 15, 2003. http://news.ft.com/world/indepth

[42] *"The Massacre of Rashdiya"*, Testimony of an Iraqi Doctor, by E.A.Khammas, Occupation Watch Center, July 28th, 2003.
http://www.occupationwatch.org/article.php?id=345

firebombs being dropped on bridges and "unfortunately, there were people there because you could see them in the (cockpit) video."[43] During the war, Pentagon spokesmen had disputed reports that napalm was being used, claiming that the Pentagon's stockpile had been destroyed two years ago. Apparently, the spokesmen were drawing a distinction between the terms "firebomb" and "napalm". If reporters had asked about firebombs, officials said they would have confirmed their use. What the Marines dropped, the spokesmen said, were "Mark 77 firebombs". They acknowledged those are incendiary devices with a function "remarkably similar" to napalm weapons. Robert Musil, executive director of Physicians for Social Responsibility, described the Pentagon's distinction between napalm and Mark 77 firebombs as "pretty outrageous. That's clearly Orwellian."

The Pentagon could not disclaim the use of depleted uranium[44]. Experts there and at the United Nations have estimated that 1,100-2,200 tons of depleted uranium were used by U.S.-led coalition forces during their attack on Iraq in March and April 2003. This contrasts with about 375 tons used in the 1991 Gulf War, 11 tons fired during the 1999 war against Serbia over Kosovo and a much smaller quantity used against rebel Serb positions in Bosnia in 1995[45].

Then, "there was no real planning for postwar Iraq," said a former senior US official who left government recently. "We could have done so much better", lamented a former senior Pentagon official, who is still a Defense Department

---

[43] *"Officials Confirm Dropping Firebombs on Iraqi Troops", Results are 'remarkably similar' to using napalm,* by James W. Crawley, The San Diego Union-Tribune, Tuesday 05 August 2003. http://www.signonsandiego.com/news/military/20030805-9999_1n5bomb.html

[44] *"Death By Slow Burn: How America Nukes Its Own Troops",* by Amy Worthington, The Idaho Observer April 16, 2003. http://proliberty.com/observer/20030401.htm

[45] *"Depleted Uranium Arms May Pose Risks",* Associated Press, June 14, 2003. http://www.khilafah.com/home/category.php?DocumentID=7524&TagID=7

adviser[46]. To retired Col. Pat Lang, who served as the Pentagon's chief of Middle Eastern intelligence from 1985 until 1992 and who has closely followed the discussions over the Iraq war and its aftermath, "It was a massive illusion that the neocons had. It all flows from that."[47]

"The Ugly American"[48] truth is that an occupation of a sovereign country, especially Iraq and its deeply entrenched roots, is not an exercise in management but a criminal act.

Israel's pre-emptive destruction of the French built research reactors to abort what it perceived as a potential nuclear threat to its security had immediately metamorphosed into a concrete effort by Iraq to obtain its nuclear weapon.

The United States' pre-emptive occupation of Iraq to abort its supposed terrorist links and weapons of mass destruction, both of which have been proven to be unsubstantiated, will see to it that the Lion of Babylon rises again.

The neoconservatives have indeed succeeded in manipulating the "American Way of Life" to devour my beloved Iraq.

We shall, however, resurrect, to their detriment.

---

[46] *"No real planning for postwar Iraq"*, by Jonathan S. Landay and Warren P. Strobel, Knight Ridder Newspapers, July 11, 2003. http://www.realcities.com/mld/krwashington/6285265.htm

[47] *"From heroes to targets"*, by Michelle Goldberg, salon.com, July 18, 2003. http://www.salon.com/news/feature/2003/07/18/pre_war/index_np.html

[48] *"The Ugly American"* by William Ledere and Eugene Burdick, 1958.

INDEX

APPENDIX

A register of the most important names in the book
and their role

Saviors and confidants

Abu Diyar, *Brave and noble savior*

Khalid Rashid, *Steadfast and most valuable support, Security officer at MIC*

Subhi Ayoub, *A trusted Jordanian friend*

Peter Kouwenburg, *Deputy Head of UNDP in Baghdad*

Fellow scientists

Jafar Dhia Jafar, *Scientific Head of the Iraqi nuclear weapons program*

Yehya al-Meshad, *Eminent Egyptian reactor physicist*

Dhafir Selbi, *Head of Group 3 and played a major role in catapulting the nuclear weapons program into high gear after 1987*

Khalid Said, *Head of the atomic bomb design of Group 4*

Basil al-Qaisi, *Senior electronics engineer and my conduit for joining the Iraqi Atomic Energy Commission*

Munqith al-Qaisi, *Senior mechanical engineer*

Mahir Sarsam, *Senior physicist in Group 4*

Sabah Abdul Noor, *Senior solid state physicist in Group 4*

Zaghloul Kassab, *Senior communications engineer in Group 4*

Hussain al-Shahrastani, *Senior chemist*

Administrative heads

Humam Abdul Khaliq, *Head of the Iraqi Atomic Energy Commission and the nuclear weapons program in the eighties*

Abdul Razzak al-Hashimi, *Inept head of the Iraqi Atomic Energy Commission in the seventies*

Mohammad al-Sahhaf, *Foreign Minister and staid supporter*

Amer al-Ubaidi, *Head of the Military Industrialization Corporation (MIC)*

Abdalla Auda, *Inept head of UNDP in Baghdad in 1998*

Colleagues and friends

Salam Toma, *A most trusted companion and confidant*

Husam Obaid, *A faithful and pure spirit*

Omran Mousa, *Faithful and reliable driver*

Adil Fiadh, *The competent head of the purchasing department in PC3*

Ayad Muhaimid, *Ever pleasant and capable engineer*

Dhia al-Taie, *Bountifully warm and supportive*

Intelligence and Security

Abu Muhannad, *An enigma and nemesis*

Abid Himood, *Secretary of Saddam and head of a suspected super intelligence organization*

Salah Abdul Rahman al-Hadithi, *An epitome of Intelligence wickedness*

Rafi' al-Dahham, *Head of Intelligence Agency in 1998*

Mohammad al-Douri (Abu Omar), *Close friend of al-Sahhaf and extended appreciated support*

Walid, *Security officer at MIC who managed to convince me to leave PC3 and the Centre for Specialized Information*

Political mentors

Adil Abid al-Mahdi, *Political intellectual and leader*

Hassan Cherif, *The conduit to the Palestine Liberation Organization and Fattah*

Sites

Nuclear Research Centre, *At Tuwaitha, also called al-Asil (the Genuine)*

Al-Safa (the Tranquil), *At Tarmiyah, the uranium enrichment plant*

Al-Fajir (the Dawn), *At Sharqat, the replica al-Safa*

Al-Athir (the Ether), *The atomic weapon design centre south west of Baghdad*

Al-Jazira (the Island), *The uranium processing plant near Mosul*